NUTRITIONAL FISH AND SHRIMP PATHOLOGY

NUTRITIONAL FISH & SHRIMP PATHOLOGY

A Handbook

European Association of Fish Pathologists (EAFP) / 5m Books Series

Albert G.J. Tacon and Loc Tran

Series Editor
Sean Monaghan

5m Books

First published 2022

Published by
5M Books Ltd
Lings, Great Easton,
Essex CM6 2HH, UK,
Tel: +44 (0)330 1333 580
www.5mbooks.com

A Catalogue record for this book is available from the British Library

ISBN 9781789181319
eISBN 9781789182231
DOI 10.52517/9781789182231

Book layout by Cheshire Typesetting Ltd, Cuddington, Cheshire
Printed by CPI Anthony Rowe Ltd, UK
Photos by authors unless otherwise indicated
Cover images: shrimp melanisation,
Dr. Amina Moss, Institute of Aquaculture, University of Stirling, Stirling, UK; fish exophthalmia & fish visceral pathology, Dr. Johanna Baily, Institute of Aquaculture, University of Stirling, Stirling, UK

Contents

Aquatic Animal Diseases Series

EAFP book series editor

Dr Sean Monaghan, EAFP Publications Officer,
University of Stirling, Scotland, UK
s.j.monaghan@stir.ac.uk

When culturing aquatic species, the proper maintenance of health is essential to the sustainability of production. Disease epidemics, for example outbreaks of infectious salmon anaemia (ISA) in Chile, and early mortality syndrome (EMS) in shrimp in South-East Asia have caused widespread and devastating effects on the aquaculture industries in the affected countries. In order for aquatic animal diseases to be properly prevented, identified, monitored, treated and managed, there is a constant need for high quality and authoritative books to inform the range of personnel working in the aquaculture industry.

The new EAFP/5m Books Series encompasses professional-level reference books and practical guides aimed at fish farm operatives and technicians, fish biologists and pathologists, veterinarians, students who are studying aquaculture and fish health, and the many personnel involved in supplying the aquaculture industry with pharmaceuticals, nutraceuticals, feed and equipment.

The first tiles in the series include:

Ariadna Sitjà-Bobadilla, James E. Bron, Geert F. Wiegertjes and M. Carla Piazzon (eds) (2021) *Fish Parasites: A Handbook of Protocols for their Isolation, Culture and Transmission*

Albert G.J. Tacon and Loc Tran (2022) *Nutritional Fish & Shrimp Pathology: A Handbook*
Bartolomeo Gorgoglione and Christyn Bailey (eds) (2022) *Proliferative Kidney Disease*

The European Association of Fish Pathologists (EAFP) exists to promote the exchange of knowledge and coordination of research related to fish and shellfish pathology. Membership of the EAFP is open to all fish and shellfish pathologists, students of fish pathology, and organisations, agencies or other associations which have an interest in maintaining the objectives of the EAFP. https://eafp.org

5m Books publishes professional-level books for people and businesses in aquaculture, fisheries, agriculture, veterinary sciences and the global food chain. www.5mbooks.com

Terminology

abscess	circumscribed area filled with liquefied cell debris and leukocytes
acute	a disease with sudden onset of signs and a short course
agglutination	clumping together of cells or particles
aggregation	a total or coming together of separate parts
anaplasia	loss of structural differentiation of cell
anemia	a condition characterized by a deficiency of hemoglobin or red blood cells
aniosocytosis	having red blood cells that are unequal in size
anorexia	diminished appetite
apoptosis	programmed cell death
ascites	accumulation of serous fluid in the abdominal cavity
ataxia	failure of muscle coordination
atrophy	a wasting away
autolysis	self-digestion by own enzymes
calculi	concretion formed by salts of inorganic or organic acids
carcinogen	a substance that causes cancer
carcinoma	malignant neoplasm of epithelial tissue
cataract	loss of transparency of eye lens or lens capsule
caudal	situated toward or pertaining to the tail

ceroid	brown pigment produced by autoxidation of unsaturated fats
cholangioma	neoplasm of bile ducts
chondroma	benign neoplasm of cartilage-forming cells
chronic	a condition with slow onset, mild but continuous manifestations and long-lasting
cirrhosis	fibrosis of liver with loss of architecture and function
clubbing	proliferation of soft tissue about the ends (terminal phalanges) of the gills
complement	factors present in the serum of normal animals which enter into various immunologic reactions
concretions	aggregation or formation of solid materials
congestion	abnormal amount of blood in vessels
cyst	abnormal sac filled with fluid or foreign material
degenerative	progressive and often irreversible deterioration
dysplasia	abnormal development of cells within tissues or organs
edema	the accumulation of excess fluid in the intercellular or interstitial tissue spaces or body cavities
encephalitis	inflammation of the brain
endocarditis	inflammation of the lining of the heart
endogenous	originating in the cells or tissues of the body
enteritis	inflammation of the intestine
epithelioma	neoplasm of epithelial cells
erythrocyte	red blood cell
etiology	the science dealing with the causes of disease
exogenous	originating from outside of the body
exophthalmia	protrusion of the eyeballs
exudate	fluid and/or blood cells pasted into tissue due to increased vascular permeability
fibroblast	elongate cell capable of forming collagen fibers

fibroma	benign neoplasm of fibrous connective tissue
fibrosis	formation of fibrous tissue in reaction to irritant
gastritis	inflammation of the stomach
gastroenteritis	inflammation of the lining of the stomach and intestine
glomerulonephritis	inflammation of glomeruli
glomerulosclerosis	hyaline deposits (scarring) in glomeruli (degenerative process)
granuloma	nodular inflammatory lesion
granulomatosis	disease characterized by multiple granulomas
heart myopathy	a disease of the heart muscle
hematocrit	volumetric relationship of the cellular elements of blood to the total blood volume
hemoglobin	the oxygen carrying pigment of the red blood cells
hemolysis	the liberation of hemoglobin or destruction of erythrocytes
hemolytic anemia	disorder in which red blood cells are destroyed faster than they can be made
hemorrhage	to bleed (an escape of blood from the blood vessels)
hemosiderin	yellow-brown pigment derived from hemoglobin breakdown
hepatitis	inflammation of the liver
hepatoma	liver cell carcinoma (malignant)
hydrocephalus	is an abnormal accumulation of fluid within cavities in the brain
hyperchromic anemia	an increase of hemoglobin in red blood cells and reduction in the number of red blood cells
hyperemia	an excess of blood in the vessels supplying an organ or other part of the body
hyperplasia	increase in number of cells in tissue or organ
hypertrophy	increase in bulk of organ or part of body

hypoplasia	decrease in number of cells (or loss of elements of tissue)
hypotrophy	decrease in bulk of organ or part of body (atrophy)
hypoxia	reduced supply of oxygen to tissues (below normal physiologic levels)
immunity	resistance to disease
infiltration	the diffusion or accumulation in a tissue or cells of substances abnormal to it
inflammation	response of body tissues to infection or injury
jaundice	yellowness of the skin, sclera, mucous membranes and excretions
lesion	wound, injury, pathologic change in tissue
leukocyte	white blood cell
leukocytosis	a transient increase in the number of white blood cells (leukocytes)
liquefaction	produces liquefied material, cells dissolved
lordosis	dorso-ventral curvature of the body (spine)
lysozyme	an enzyme capable of destroying certain bacterial cell walls
macrophage	mononuclear phagocytic cell
malignant	(neoplasm) having uncontrolled growth and dissemination through the body
melanin	brown to black pigment derived from tyrosine by melanocytes
melanophore	melanin-containing cell contributing to color changes in fish
melanoma	malignant neoplasm derived from melanocytes
microcytic anemia	is a condition in which the body's tissues and organs do not get enough oxygen
muscular dystrophy	group of disorders leading to a progressive weakening and breakdown of skeletal muscles
myositis	inflammation of muscle

necrosis	death of one or more cells
neoplasm	abnormal mass of tissue with uncoordinated growth
nephritis	inflammation of kidney tissue except tubules (nephrosis)
nephrocalcinosis	presence of calcification in the kidney
nephrolithiasis	presence of calculi in kidney
nephron	functional unit of kidney (glomerulus, proximal and distal tubules)
nephrosis	diseases of the kidney tubules
neutrophil	a granular leukocyte having a nucleus with three to five lobes
occlusion	closing or shutting off
oxidative stress	an imbalance of free radicals and antioxidants in the body which can lead to cell and tissue damage
pericarditis	inflammation of the pericardium (sac enclosing the heart)
poikilocytosis	is the term for having abnormally shaped red blood cells
scoliosis	lateral curvature of the body (spine)
spasm	a sudden, violent, involuntary muscle contraction
splenomegaly	enlargement of the spleen
steatosis	fatty degeneration
stroma	the connective tissue framework of an organ or other structure
subcutaneous	beneath the skin
telangiectasia	dilation of terminal blood vessels (spaces)
teratogen	a substance or condition that impairs normal development of the embryo or fetus in utero
tetany	involuntary muscle contraction & state of hyperexcitability, muscle cramps and spasms
therapeutic	serving to heal or cure
ulcer	an open sore or break in the skin or a mucous membrane

1

Introduction

Genetics and nutrition determine the health and well-being of all animal species, including fish and shrimp. Since nutrition is the study of the biological processes of growth, reproduction, and maintenance, depending upon the intake and digestion of food, and the subsequent absorption of dietary nutrients, it follows therefore that any negative alterations to any of these biological processes can result in nutritional disorders, tissue pathologies, and fish and shrimp health risks.

Of particular note, is the deleterious effect of environmental *stress* and unfavorable rearing conditions on feed intake and dietary nutrient requirements, and in particular the potential *stressful* effects of sub-optimal water quality conditions (temperature, oxygen, salinity, natural food availability, pollutants, algal blooms, weather, storm-water run-off) and sub-optimal rearing conditions (stocking density, water exchange, sediment build-up, human interactions, disease outbreaks) on the farmed species. Notwithstanding the above, it is generally believed that the majority of these environmental stress factors can be reduced and mitigated by creating a more stable culture environment and effective management system for the farmed species [2–6].

This review focuses on the major nutritional disorders reported in farmed fish and shrimp, and is an update to previous review by Tacon [7].

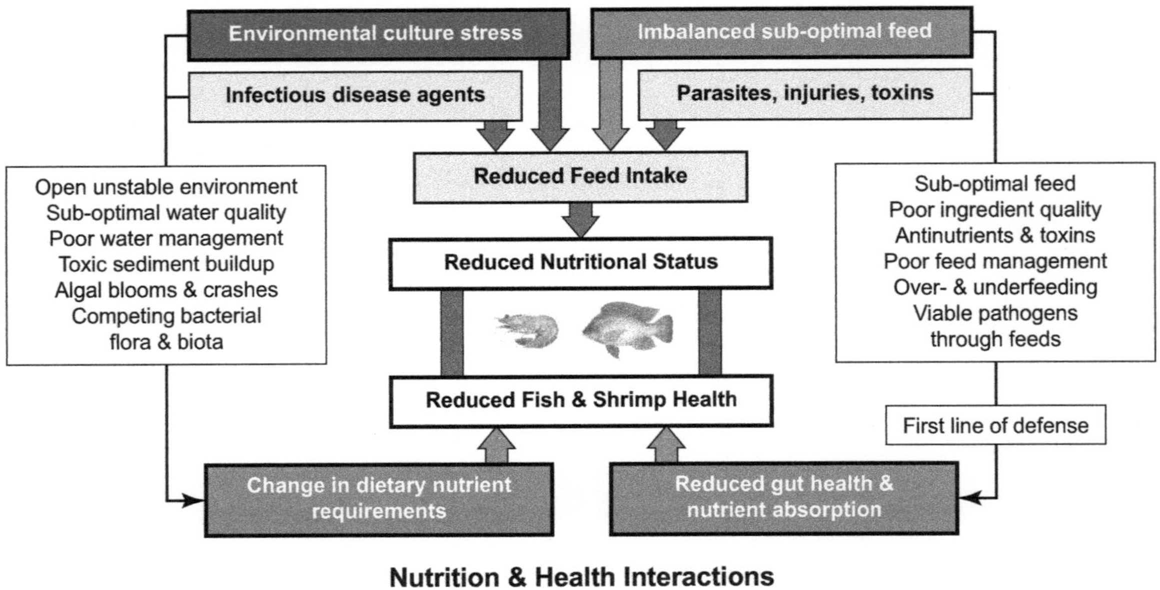

Figure 1.1 Diagrammatic representation of the possible environmental interactions between nutrition and health in farmed fish and shrimp. Modified after Waagbø [1] and Tacon [2].

2

Disorders in protein and amino acid nutrition

Proteins are composed of 20 or more amino acids in most farmed fish and shrimp species, and they represent an essential component of all living cells, including all enzymes and hormones, muscle tissue, and are fundamental for proper metabolic, neural, and digestive function. Moreover, dietary protein selection and use usually represents the largest operating cost item within most fish and shrimp feed formulations. It is perhaps not surprising therefore that in a global world of increasing feed ingredient prices and increasing competition for limited high-quality protein sources, such as fish meal, that diets are sometimes formulated and produced with sub-optimal and/or imbalanced amino acid levels; the first three limiting essential amino acids within commercially formulated fish and shrimp feeds usually being lysine, methionine and/or threonine.

Table 2.1 shows the reported deficiency signs for individual amino acids when present at sub-optimal levels (that is, below known dietary requirement levels) within formulated diets for different farmed fish and shrimp species. Of particular significance from the data presented in Table 2.1, is the important metabolic role played by many amino acids on fish and shrimp health, immunity, gut health, and disease resistance, and in particular the functional amino acids arginine, histidine, isoleucine, methionine, phenylalanine, taurine, threonine, tryptophan, and valine.

Although specific dietary amino acid deficiencies can be remedied through dietary supplementation with the corresponding limiting free amino acid [94], it is also apparent from the deficiency

Table 2.1 Reported amino acid deficiency signs and health impacts in farmed fish and shrimp.

Major species by amino acid	Reported amino acid deficiency signs and health impacts[a]
Arginine	
Channel catfish *Ictalurus puctatus*	– Reduced disease resistance & immune response [8–10]
Golden pompano *Trachinotus ovatus*	– Reduced disease resistance & immune response [11]
Jian carp *Cyprinus carpio var.* Jian	– Reduced disease resistance & impaired intestinal immune response [12–13]
Largemouth bass *Micropterus salmoides*	– Reduced immune response [14]
Nile tilapia *Oreochromis niloticus*	– Reduced disease resistance & immune response [15–16]
Red drum *Sciaenops ocellatus*	– Reduced immune response & intestinal structure [17]
Red sea bream *Pagrus major*	– Reduced immune response [18]
Senegalese sole *Solea senegalensis*	– Reduced disease resistance & immune response [19]
Striped bass *Morone chrysops* × *saxatilis*	– Reduced immune response & intestinal functionality [20]
Sturgeon *Acipenser schrrenckii* × *baeri*	– Reduced stress control, disease resistance & immune response [21–22]
Turbot *Scophthalmus maximus*	– Reduced digestive capacity & intestinal morphology [23]
Yellow catfish *Pelteobagrus fulvidraco*	– Reduced disease resistance & immune response [24]
Histidine	
Atlantic salmon *Salmo salar*	– Cataract [85, 88, 91–92]
Blunt snout bream *Megalobrama amblycephala*	– Reduced intestinal health & antioxidant capacity [84]
Giant tiger prawn *Penaeus monodon*	– Abnormal histological alterations in the hepatopancreas [93]
Grass carp *Ctenopharyngodon idella*	– Reduced gill structural integrity & antioxidant capacity [86–87]
Jian carp *Cyprinus carpio* var. Jian	– Reduced digestive & antioxidant capacity [89–90]
Isoleucine	
Jian carp *Cyprinus carpio* var. Jian	– Reduced intestinal immune response, tight junctions & microflora [25–26]
Olive flounder *Paralichthys olivaceus*	– Reduced innate immunity [27]

Lysine	
Atlantic salmon *Salmo salar*	– Loss of appetite [33]
Black seabream *Sparus macrocephalus*	– Reduced hemoglobin level [30]
Cobia *Rachycentron canadum*	– Reduced hematocrit [31]
Milkfish *Chanos chanos*	– Reduced appetite [34]
Rainbow trout *Oncorhynchus mykiss*	– Caudal fin erosion & increased mortality in fry [35]
Silver perch *Bidyanus bidyanus*	– Reduced hematocrit & hemoglobin concentration [29]
Yellow Perch *Perca flavescens*	– Reduced sperm quality & motility [28]
Yellowtail *Seriola quinqueradiata*	– Reduced hematocrit [32]
Methionine	
Arctic charr *Salvelinus alpinus*	– Cataract [40]
Atlantic salmon *Salmo salar* L.	– Loss of appetite [41]
European seabass *Dicentrarchus labrax*	– Reduced innate immune response [38]
Jian carp *Cyprinus carpio* var. Jian	– Reduced digestive capacity, microflora & immune response [39]
Nile tilapia *Oreochromis niloticus*	– Reduce immune response [36]
Yellow catfish *Pelteobagrus fulvidraco*	– Reduced disease resistance & immune response [37]
Phenylalanine	
Grass carp *Ctenopharyngodon idella*	– Reduced digestive & immune function & antioxidant capacity [42–43]
Taurine[b]	
European seabass *Dicentrachus labrax*	– Reduced immune function [71]
Hybrid snakehead *Channa maculatus* × *C. argus*	– Reduced stress resistance [72]
Japanese flounder *Paralichthys olivaceus*	– Reduced oxidative stress response [73]
Rabbitfish *Siganus canaliculatus*	– Reduced stress response [74]
Red sea bream *Pagrus major*	– Green liver, increased scale loss [75–78]
Tongue sole *Cynoglossus semilaevis*	– Reduced reproductive performance, antioxidant function & larval quality [79]
Totoaba *Totoaba macdonaldi*	– Green liver, low visceral fat & fat digestibility [80]
Turbot *Scophthalmus maximus*	– Reduced immune function [81]
Yellow catfish *Pelteobagrus fulvidraco*	– Reduced immune function, hemoglobin & hematocrit [82]
Yellowtail *Seriola quinqueradiata*	– Green liver [83]

Table 2.1 (Continued)

Major species by amino acid	Reported amino acid deficiency signs and health impacts[a]
Threonine	
Blunt snout bream *Megalobrama amblycephala*	– Reduced immune response, intestinal function & hemoglobin [44–46]
Grass carp *Ctenopharyngodon idella*	– Reduced intestinal morphology & function [47]
Tryptophan	
Asian seabass *Lates calcarifer*	– Increased cannibalism in fry [51]
Ayu *Plecoglossus altivelis*	– Reduced maturation [65]
Atlantic cod *Gadus morhua*	– Increased aggressive behaviour & reduced stress response [59, 62]
Brycon *Brycon amazonicus*	– Altered aggressive behaviour [57]
Chum salmon *Oncorhynchus keta*	– Cataract, scoliosis [67]
Common carp *Cyprinus carpio*	– Reduced stress response [60]
European seabass *Dicentrarchus labrax*	– Reduced immune response [56]
Giant tiger prawn *Penaeus monodon*	– Caudal fin erosion [66]
Grass carp *Ctenopharyngodon idellus*	– Gill antioxidant system disruption [55]
Grouper *Epinephelus coioides*	– Increased cannibalism [63]
Hybrid catfish *Pelteobagrus vachelli* × *L. longirostris*	– Reduced appetite, digestive function & antioxidant status [49]
Jian carp *Cyprinus carpio* var. Jian	– Reduced digestive function [58]
Mrigal carp *Cirrhinus mrigala*	– Reduced stress response [61]
Persian sturgeon *Acipenser persicus*	– Reduced stress & immune response [52]
Pikeperch *Sander lucioperca*	– Increased cannibalism in fry [53]
Rainbow trout *Oncorhynchus mykiss*	– Reduced disease resistance, stress & immune response [48, 64]
Totoaba *Totoaba macdonaldi*	– Reduced stress response [50]
White shrimp *Litopenaeus vannamei*	– Reduced protein digestibility & protease activity [54]
Valine	
Grass carp *Ctenopharyngodon idella*	– Impaired gill structural integrity & gill antioxidant status [68]
Jian carp *Cyprinus carpio var. Jian*	– Reduced digestive function & microbiota [70]
Red sea bream *Pagrus major*	– Reduced immune response [69]

Notes: [a]Major reported deficiency signs and health impacts in addition to reduced growth and feed efficiency; [b]non-essential amino acid for most non-carnivorous fish species.

signs reported in Table 2.1, that many amino acids perform other important biological functions over and above that of just serving as building blocks for proteins and tissue growth (Figure 2.1; [95]). It follows, therefore, that dietary amino acid requirements may be higher when formulating feeds for optimum health and disease resistance, gut health and immunity, stress control, and reproductive function [96–98].

Compared to fishmeal, which has a well balanced amino acid profile, most conventional terrestrial animal by-product meals, and plant oilseed and cereal by-product meals have one or more specific amino acid deficiencies and/or imbalances that prevent them from being used as the sole source of dietary protein for most fish species without supplementation with limiting free amino acids. Table 2.2 shows the calculated essential amino acid (EAA) score of 94 different animal and plant feed ingredients compared with the estimated EAA requirement of fish [99]. From the data presented, methionine was found to be the first limiting EAA (50 ingredients having a

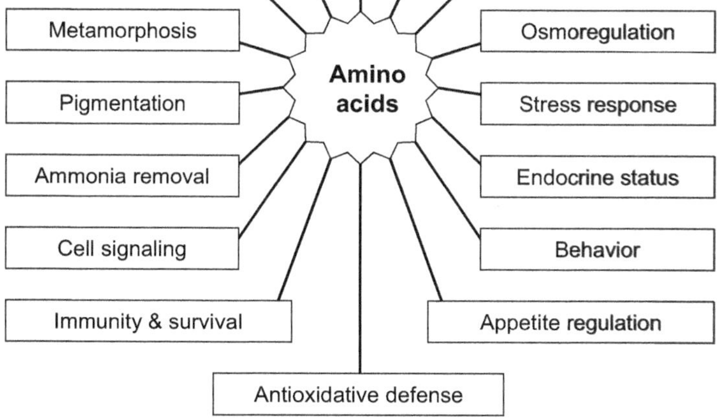

Figure 2.1 Different roles played by amino acids in the growth, development and health of fish. Adapted after Li et al. [95].

Table 2.2 Calculated essential amino acid (EAA) score[a] of different animal and plant feed ingredients compared with the estimated EAA requirement profile of fish[b]

Product/EAA	Arg	Cys	Met	Thr	Iso	Leu	Lys	Val	Tyr	Try	Phe	His	Diff[c]
Ingredient EAA profile compared with estimated dietary EAA requirements of fish													
Fishery products													
Anchovy fishmeal	98	70	109	80	123	112	91	112	103	135	86	100	65
Menhaden fishmeal	104	67	107	76	125	112	93	113	98	129	84	98	62
Herring fishmeal	107	74	109	78	117	110	91	122	95	129	81	98	55
Tuna fishmeal	109	63	100	81	120	104	91	108	97	123	84	135	72
White fishmeal	120	81	107	81	117	109	92	106	95	123	80	94	43
Fish solubles, condensed	108	74	105	66	105	112	87	106	51	135	81	244	193
Fish solubles, dehydrated	110	104	91	79	115	100	92	98	61	276	73	177	215
Shrimp meal (process residue)	144	93	80	80	121	106	68	107	100	106	103	100	76
Shrimp head meal	104	63	163	73	101	92	70	103	123	123	142	90	100
Krill meal	106	93	107	76	128	108	86	102	118	129	94	85	53
Squat lobster/langostilla meal	61	74	70	78	112	96	78	117	148	159	91	169	108
Squid meal	109	104	83	80	105	103	80	98	126	188	88	154	108
Squid liver meal	105	89	87	73	132	95	71	95	95	259	89	206	188
Terrestrial livestock products													
Blood meal, spray dried	62	63	30	71	25	167	86	160	69	135	127	187	152
Feather meal, hydrolysed	131	322	26	90	128	129	31	159	91	88	88	42	296
Meat and bone meal, rendered	141	93	65	78	103	115	79	122	88	100	90	106	76
Meat meal, rendered	144	111	63	73	109	112	87	123	68	106	88	96	81
Poultry by-product meal, rendered	134	130	78	78	120	120	72	117	80	106	87	98	62
Liver meal	104	104	68	71	124	118	93	132	78	123	93	92	56

Terrestrial invertebrate products													
Soldier fly larvae	83	16	73	21	117	113	89	157	169	51	102	173	157
Earthworm meal	141	93	65	78	103	115	79	122	88	100	90	106	76
Marine polychaete meal	140	78	48	94	119	120	75	110	112	94	82	87	92
Protein-rich cereal products													
Brewer's grains (5–02–141)	82	96	63	67	157	144	40	127	143	165	117	83	125
Malt sprouts (culms)	95	89	59	89	136	113	69	143	89	229	90	102	170
Maize gluten feed	81	189	91	80	100	163	37	120	123	105	91	142	152
Maize gluten meal (41% protein)	54	118	87	62	132	238	21	103	83	59	136	90	217
Maize gluten meal (60% protein)	53	126	94	61	103	232	19	96	155	65	127	83	213
Maize DDS	70	141	83	77	139	136	46	133	108	106	126	112	95
Maize DDG	83	100	78	37	124	202	48	121	125	118	76	123	165
Maize DDGS	75	118	78	76	145	124	34	134	103	71	129	112	111
Maize germ meal	106	189	109	102	93	115	51	125	85	129	85	146	138
Wheat germ meal	142	144	68	81	101	91	81	112	101	153	89	115	85
Wheat gluten meal	80	215	78	64	127	138	26	108	141	141	146	115	189
Rice protein concentrate	130	152	91	71	103	118	36	114	161	153	108	90	117
Oilseed protein products													
Canola meal, solvent extracted	110	148	78	87	112	108	74	111	88	153	93	133	79
Canola protein concentrate	118	167	81	83	109	117	71	108	88	159	88	110	96
Rapeseed meal, mech. extracted	110	78	81	93	120	115	65	122	85	170	96	123	105
Rapeseed meal, solv. extracted	111	85	80	92	113	115	76	118	75	159	93	127	84
Coconut kernel (endosperm), dry	230	85	72	63	107	100	42	115	83	123	96	87	188
Copra meal, mech. extracted	219	104	65	67	117	108	38	116	85	123	95	79	181
Copra meal, solv. extracted	217	96	61	65	116	112	37	115	92	123	95	83	180
Cottonseed (kernel), whole	202	118	50	65	92	92	55	102	94	153	115	119	152

Table 2.2 (Continued)

Product/EAA	Arg	Cys	Met	Thr	Iso	Leu	Lys	Val	Tyr	Try	Phe	His	Diff[c]
Cottonseed meal, mech. extr.	189	141	55	66	101	92	50	110	85	165	119	112	139
Cottonseed meal, solv. extr.	197	133	55	64	97	86	56	104	77	159	126	114	142
Cottonseed meal (dehul. solv. extr.)	196	181	67	74	93	80	48	107	58	171	130	119	148
Linseed/flax (kernel)	171	148	76	74	120	94	47	118	88	188	105	90	141
Linseed meal, mech. extr.	163	137	68	73	148	95	46	115	92	200	99	87	154
Linseed meal, solv. extr.	159	141	61	74	147	95	43	115	108	188	99	92	145
Oil palm seed (kernel)	239	133	89	61	53	92	43	118	85	112	80	90	196
Oil palm (kernel), solv. extracted	238	122	72	67	100	103	37	101	85	135	97	77	201
Peanut meal, mech. extracted	204	130	43	55	105	104	42	102	120	129	115	104	162
Peanut meal, solvent extracted	201	115	37	59	111	103	48	96	125	129	109	106	164
Safflower seed (kernel)	179	126	43	69	132	110	45	123	88	135	90	119	136
Safflower oilcake decort. mech. extr.	139	152	85	74	124	105	36	131	85	235	107	110	199
Safflower oilmeal decort. solv. extr.	170	141	68	67	116	98	41	132	88	188	102	117	147
Sesame seed (kernel)	212	137	105	68	97	101	33	98	98	159	95	102	179
Sesame oilcake, mech. extr.	194	133	107	68	108	102	31	101	117	159	96	100	163
Soybean seed, heat processed	123	93	50	79	139	104	74	112	97	159	109	104	109
Soybean meal, undehull. mech. extr.	117	93	52	71	156	118	72	104	105	159	100	102	107
Soybean meal, undehull. solv. extr.	131	122	48	74	127	112	76	101	100	171	102	108	123
Soybean meal, dehulled, solv. extr.	128	111	52	73	124	110	75	107	108	159	105	102	107
Soybean protein concentrate	131	100	54	74	128	110	74	104	106	141	104	110	87
Sunflower seed (kernel) with hulls	155	122	78	78	129	107	49	119	65	176	105	110	127
Sunflower seed, decort. solv. extr.	153	118	91	74	127	107	48	120	81	153	105	102	105

Pulse/grain legume seed products													
Pigeon pea	93	78	22	61	93	105	103	84	69	76	196	175	174
Jack bean	89	96	54	91	117	118	72	106	118	153	119	123	99
Chickpea	159	85	37	70	116	110	80	94	89	100	118	108	122
Egyptian bean	116	74	22	67	117	126	90	106	105	88	108	133	111
Lentil	147	67	28	74	113	112	84	104	98	112	108	112	119
Lupin	174	107	30	74	124	113	67	90	115	123	82	115	144
African locust bean	119	96	39	65	116	112	84	109	121	118	109	123	84
Lima bean	99	74	46	76	128	117	86	104	97	118	123	127	82
Kidney bean	103	74	43	80	120	120	92	103	83	129	117	127	86
Pea/field pea	123	67	41	77	127	113	78	109	106	106	109	112	86
Pea protein concentrate	162	89	30	68	112	30	87	97	95	112	105	106	132
Urd	98	55	20	65	217	108	99	89	80	94	106	125	197
Broad bean	160	52	35	74	115	108	80	106	112	118	94	108	125
Cowpea	119	85	46	73	109	112	87	102	86	135	117	146	100
Bambarra groundnut	109	74	65	66	117	116	77	113	108	129	119	125	64
Ground beans	110	70	52	72	119	112	78	130	106	88	120	115	78
Miscellaneous plant protein products													
Leaf protein concentrate	100	59	59	84	119	126	70	122	112	159	118	85	100
Potato protein concentrate	69	111	83	86	135	121	77	102	158	118	114	62	96
Cassava leaf meal	83	81	48	81	100	136	112	119	111	118	75	94	88
Alfalfa leaf meal (22% protein)	81	107	61	90	139	118	56	133	97	282	117	90	226
Single cell proteins													
Pseudomonas/*Methylophilus* spp.	92	44	94	90	129	117	74	127	120	135	94	92	91
Mixed bacterial SCP	121	52	102	85	119	115	67	122	109	159	87	92	107
Brewer's yeast (*S. cerevisiae*)	89	92	65	95	133	107	87	118	108	153	88	108	98
Extracted yeast (*S. cerevisiae*)	80	93	67	89	125	129	79	125	111	141	95	98	74

Table 2.2 (Continued)

Product/EAA	Arg	Cys	Met	Thr	Iso	Leu	Lys	Val	Tyr	Try	Phe	His	Diff[c]
Torula yeast (*T. utilis*)	85	85	54	93	144	98	84	117	117	112	115	106	90
Candida spp. (alkane substrate)	73	85	76	103	136	118	80	120	118	165	89	85	89
Aspergillus oryzae (waste starch)	97	74	54	88	121	112	68	117	195	171	75	102	141
Rhodotorula pilimanae	129	18	93	101	113	104	104	113	83	35	72	110	111
Spirulina maxima	113	41	65	87	159	121	55	129	131	165	92	69	124
Activated bacterial SCP (brewery)	92	59	78	88	135	112	77	110	129	171	103	90	112
Ingredient EAA profile compared with estimated dietary EAA requirements of shrimp													
Fishery products													
Anchovy fishmeal	59	95	140	106	112	106	143	116	82	128	83	112	84
Menhaden fishmeal	63	90	138	101	115	106	146	118	78	122	81	109	83
Herring fishmeal	64	100	140	104	107	104	143	127	76	122	78	109	77
Tuna fishmeal	66	85	129	107	110	99	143	113	77	117	81	151	85
White fishmeal	72	110	138	107	107	103	145	111	76	117	77	105	67
Fish solubles, condensed	65	100	136	87	96	106	136	111	40	128	78	272	232
Fish solubles, dehydrated	66	140	117	105	105	95	145	102	49	261	70	198	212
Shrimp meal (process residue)	86	125	102	106	111	101	107	112	79	100	99	112	46
Shrimp head meal	63	85	209	96	93	87	110	108	98	117	136	100	146
Krill meal	64	122	137	102	117	103	135	107	94	121	90	95	73
Squat lobster/langostilla meal	58	100	90	104	102	91	122	122	117	150	87	188	130
Squid meal	65	140	107	106	96	98	125	102	100	178	85	172	113
Squid liver meal	63	120	112	96	121	91	112	99	76	244	86	230	181

Terrestrial livestock products													
Blood meal, spray dried	37	83	38	94	23	158	135	167	55	127	122	209	186
Feather meal, hydrolysed	79	424	33	119	117	122	49	166	72	83	85	46	391
Meat and bone meal, rendered	85	122	82	104	93	110	125	127	69	94	87	119	58
Meat meal, rendered	86	146	80	97	100	106	138	129	53	100	85	107	93
Poultry by-product meal, rendered	80	171	99	104	109	114	114	122	63	99	84	109	108
Liver meal	63	137	87	94	113	113	147	137	62	116	89	102	84
Terrestrial invertebrate products													
Soldier fly larvae	50	21	93	27	107	108	140	164	134	48	97	194	173
Earthworm meal	85	122	82	104	93	110	125	127	69	94	87	119	58
Marine polychaete meal	84	102	61	125	108	114	118	115	89	88	79	98	64
Protein-rich cereal products													
Brewer's grains (5–02–141)	49	127	80	89	143	137	64	133	113	155	112	93	106
Malt sprouts (culms)	57	117	75	118	124	108	109	149	70	215	87	114	158
Maize gluten feed	49	249	116	107	91	155	59	125	97	99	87	158	200
Maize gluten meal (41% protein)	33	156	111	83	120	227	34	108	66	56	130	100	194
Maize gluten meal (60% protein)	32	166	120	82	93	221	30	100	123	61	122	93	191
Maize DDS	42	185	106	103	126	130	72	138	85	99	121	126	143
Maize DDG	50	132	99	49	113	192	76	126	98	110	73	137	143
Maize DDGS	45	156	99	102	132	196	53	140	81	66	124	126	151
Maize germ meal	64	249	139	135	85	110	80	131	67	121	82	163	185
Wheat germ meal	85	190	87	108	92	87	128	116	80	144	86	128	110
Wheat gluten meal	48	283	99	85	115	131	41	113	112	133	140	128	242
Rice protein concentrate	78	200	116	94	94	113	57	119	128	144	104	100	143

Table 2.2 (Continued)

Product/EAA	Arg	Cys	Met	Thr	Iso	Leu	Lys	Val	Tyr	Try	Phe	His	Diff[c]
Oilseed protein products													
Canola meal, solvent extracted	66	195	199	115	102	103	116	115	69	144	89	149	129
Canola protein concentrate	71	219	104	110	100	111	113	113	69	149	85	123	150
Rapeseed meal, mech. extracted	66	102	104	124	109	110	102	127	67	160	92	137	94
Rapeseed meal, solv. extracted	67	112	101	123	103	110	119	123	60	149	89	142	89
Coconut kernel (endosperm) dry	138	112	92	84	97	95	66	120	66	116	92	98	72
Copra meal, mech. extracted	131	136	82	89	107	103	60	121	67	116	91	88	76
Copra meal, solv. extracted	130	127	78	87	106	106	59	120	73	116	91	93	71
Cottonseed (kernel), whole	122	156	64	87	84	87	86	106	74	144	110	132	92
Cottonseed meal, mech. extr.	113	185	71	88	92	88	79	115	67	155	114	126	118
Cottonseed meal, solv. extr.	118	176	71	85	89	82	89	109	61	149	121	128	115
Cottonseed meal (dehul. solv. extr.)	118	239	85	98	88	76	75	112	46	160	125	132	193
Linseed/flax (kernel)	103	195	97	99	109	89	74	123	69	177	101	100	126
Linseed meal, mech. extracted	98	180	87	97	135	91	73	120	73	188	95	98	115
Linseed meal, solv. extracted	96	185	78	98	134	91	68	120	85	177	95	102	117
Oil palm seed (kernel)	143	176	113	81	87	87	67	123	67	105	77	100	109
Oil palm (kernel), solv. extracted	143	161	92	89	91	98	59	105	67	127	93	86	102
Peanut meal, mech. extracted	123	171	54	73	96	99	66	106	95	121	110	116	117
Peanut meal, solvent extracted	121	151	47	79	101	98	75	100	98	121	105	119	104
Safflower seed (kernel)	108	166	54	92	120	105	70	129	69	127	87	132	112
Safflower oilcake, decort. mech. extr.	83	200	108	99	113	100	57	137	67	221	103	123	164
Safflower oilmeal decort. solv. extr.	102	185	87	89	106	94	65	137	69	177	98	130	120
Sesame seed (kernel)	127	180	134	90	89	96	52	102	78	149	91	114	128
Sesame oilcake, mech. extracted	116	176	137	90	98	97	49	105	92	149	92	112	127

Soybean seed, heat processed	74	122	64	105	126	99	116	116	76	149	105	116	85
Soybean meal, undehull. mech. extr.	70	122	66	94	142	113	113	109	83	149	96	114	83
Soybean meal, undehull. solv. extr.	79	161	61	98	115	106	119	105	79	160	98	121	100
Soybean meal, dehulled, solv. extr.	77	146	66	97	113	105	118	112	85	149	101	114	80
Soybean protein concentrate	79	132	68	99	117	104	117	109	84	133	100	123	65
Sunflower seed (kernel) with hulls	93	161	99	104	118	101	77	124	51	166	101	123	115
Sunflower seed, undec. solv. extr.	102	200	99	110	104	96	82	138	–	177	98	144	118
Sunflower seed, decor. mech. extr.	97	180	130	93	120	91	80	119	64	149	101	116	116
Sunflower seed, decor. solv. extr.	92	156	115	98	115	102	75	125	64	144	101	114	92
Pulse/grain legume seed products													
Pigeon pea	56	102	28	81	85	100	162	88	55	72	188	195	166
Jack bean	53	127	68	122	107	112	113	111	93	144	114	137	91
Chickpea	96	112	47	93	106	104	127	98	70	94	113	121	80
Egyptian bean	70	97	28	89	107	120	142	111	83	83	104	149	121
Lentil	88	88	35	98	103	106	133	109	78	105	104	126	98
Lupin	104	141	38	98	113	108	105	94	91	116	98	128	103
African locust bean	71	127	49	87	106	106	133	114	96	110	105	137	88
Lima bean	59	97	59	102	117	111	136	109	76	110	118	142	83
Kidney bean	62	97	54	107	109	114	144	108	66	121	112	142	90
Pea/field pea	74	88	52	103	115	107	123	114	84	99	105	126	74
Pea protein concentrate	97	117	38	90	102	38	137	101	75	105	101	119	99
Urd	59	73	26	87	198	103	157	93	63	88	102	139	172
Broad bean	96	68	45	99	104	103	127	111	89	110	90	121	82
Cowpea	71	112	59	97	100	106	138	107	68	127	112	163	104
Bambarra groundnut	66	97	82	88	107	111	122	117	85	121	114	139	73
Ground bean	66	93	66	95	108	106	123	136	84	83	115	128	62

Table 2.2 (Continued)

Product/EAA	Arg	Cys	Met	Thr	Iso	Leu	Lys	Val	Tyr	Try	Phe	His	Diff[c]
Miscellaneous plant protein products													
Leaf protein concentrate	60	78	75	112	108	120	111	127	89	149	113	95	89
Potato protein concentrate	41	146	106	114	123	115	121	107	125	110	109	70	105
Cassava leaf meal	50	107	61	108	91	129	176	124	87	110	72	105	126
Alfalfa leaf meal (22% protein)	49	141	78	119	126	112	89	138	76	265	112	100	216
Single cell proteins													
Pseudomonas/*Methylophilus* spp.	55	58	120	120	118	111	117	133	95	127	90	102	78
Mixed bacterial SCP	73	68	130	113	108	110	106	127	86	149	84	102	81
Brewer's yeast (*S. cerevisiae*)	53	122	82	127	121	102	137	123	85	144	85	121	91
Extracted yeast (*S. cerevisiae*)	48	122	85	118	114	123	124	131	87	133	91	109	85
Torula yeast (*T. utilis*)	51	112	68	124	131	94	133	122	92	105	110	119	82
Candida spp. (alkane substrate)	44	112	97	137	124	113	126	125	93	155	86	95	111
Aspergillus oryzae (waste starch)	58	98	58	117	111	106	108	122	154	160	72	114	102
Rhodotorula pilimanae	78	24	118	134	103	99	163	118	66	33	69	123	139
Activated bacterial SCP (brewery)	55	78	99	117	123	106	121	115	102	160	99	100	105

Notes: arginine-Arg; cystine-Cys; methionine-Met; threonine-Thr; isoleucine-Iso; leucine-Leu; lysine-Lys; valine-Val; tyrosine-Tyr; tryptophan-Try; phenylalanine-Phe; histidine-His.

[a]Calculated essential amino acid (EAA) score calculated by comparing the level of the different EAA with the EAA requirements of farmed finfish and shrimp, respectively.

[b]EAA requirement profile of farmed fish (Arg 11.6%, Cys 2.7%, Met 5.4%, Thr 10.6%, Iso 7.5%, Leu 13.5%, Val 9.5%, Tyr 6.5%, Try 1.7%, Phe 9.5% and His 4.8%). EAA requirement profile of farmed shrimp (Arg 19.3%, Cys 2.0%, Met 4.2%, Thr 8.0%, Iso 8.2%, Leu 14.2%, Val 9.1%, Tyr 8.2%, Try 1.8%, Phe 9.9% and His 4.3%) [99].

[c]Difference between the highest and lowest calculated EAA score.

methionine level below 70% of the estimated dietary methionine requirement for fish), followed by lysine (40 ingredients), and threonine (27 ingredients).

Of particular note from the data presented, was the imbalanced EAA profile of several key animal and plant protein sources, including feather meal (Cys-322, Met-26, Lys-31, His-42), blood meal (Met-30, Iso-25, His-187), soldier fly larvae (Cys-16, Thr-21, Try-51), corn gluten meal 60% (Arg-53, Thr-61, Leu-232, Lys-19, Try-65), cottonseed meal, solvent extracted (Arg-197, Met-55, Thr-64, Lys-56), groundnut meal, solvent extracted (Arg-201, Met-37, Thr-59, Lys-48), and pea protein concentrate (Met-30, Thr-68, Leu-30). Clearly, notwithstanding the above imbalances and deficiencies, with careful selection of complementary dietary protein sources and the use of free amino acids [94], diets can be formulated to meet all the dietary EAA requirements of the cultured species at the desired level.

Under practical farming conditions, difficulties may arise from the use of supplemental free amino acids, either due to nutrient leaching on prolonged immersion in water [100–103], the differential absorption of free and protected amino acids and dipeptides [104–106], or loss of amino acid bioavailability and/or destruction during the feed extrusion process [95, 107–108].

3

Disorders in lipid nutrition

3.1 Dietary essential fatty acid deficiency

Dietary essential fatty acids (EFA) are important components of fats and oils, and may include (depending upon the fish or shrimp species): linoleic acid (18:2n-6), linolenic acid (18:3n-3), eicosapentaenoic acid (EPA 20:5n-3), docosahexaenoic acid (DHA 22:6n-3), and/or arachidonic acid (20:4n-6). Dietary EFA are defined as those fatty acids that the fish or shrimp species in question are incapable of synthesizing at a sufficient rate to meet metabolic demand. EFA have numerous important biological roles within the animal body, including serving as essential structural components of all cell membranes, intracellular signaling, hormone production, and energy production [109–110]. Table 3.1 shows the major reported deficiency signs and health impacts of fish and shrimp fed EFA deficient diets.

As with individual amino acids, the EFA also have functional properties [142], and as such apart from a basic dietary requirement that needs to be met to prevent classic nutritional pathologies affecting optimum growth and feed efficiency, there is also an additional possible requirement for maximum growth and optimum health under stressful environmental conditions [143].

Table 3.1 Reported essential fatty acid (EFA) deficiency signs and health impacts in farmed fish and shrimp.

Major species – EFA	Reported EFA deficiency signs and health impacts[a]
Atlantic salmon *Salmo salar*	Increased fry mortality [111], increased heart weight & atheriosclerotic alterations [112], altered gut microbiota [113]
Blue shrimp *Penaeus stylirostris*	Reduced resistance to environmental stressors & immune response [114]
Blunt snout bream *Megalobrama amblycephala*	Reduced immune function & antioxidant status [115]
Chum Salmon *Oncorhynchus keta*	Increased mortality, swollen pale livers [116, 141]
Dover sole *Solea solea*	Reduced larval tolerance to hypoxia [117]
European sea bass *Dicentrarchus labrax*	Reduced non-specific immune function [118]
Gilthead sea bream *Sparus aurata*	Reduced erythrocyte volume, increased erythrocyte fragility, hemoglobin count, red blood cell count, and reduced cellular immunity [119], reduced egg production and egg quality [120], Liver granulomatosis, extensive hemorrhages of internal organs, liver degeneration, pericarditis and endocarditis [121], high occurrence of hydrops in larvae [122]
Grass carp *Ctenopharyngodon idella*	Vertebral column curvature (lordosis) [123, 141]
Greasy grouper *Epinephelus tauvina & E. fuscoguttatus*	Larvae – shock syndrome [124]
Japanese seabass *Lateolabrax japonicus*	Reduced immune response [125]
Milkfish *Chanos chanos*	Increased eye abnormality in larvae [126], Increased larval mortality [127]
Nile tilapia *Oreochromis niloticus*	Swollen pale livers & high liver lipid content [128]
Palmetto bass *Morone saxatilis × M. chrysops*	Shock syndrome – larvae [129]
Rainbow trout *Oncorynchus mykiss*	Reduced reproductive performance and egg quality, including reduced eyed eggs and hatchability [130], fin erosion, heart myopathy, shock syndrome [131], Increased mortality & shock syndrome [132], Poor appetite, shock syndrome [133], reduced immune response [134], Histological alterations [135]
Red drum *Sciaenops ocellatus*	Increased mortality, fin erosion, shock syndrome [136]
Striped Jack *Pseudocaranx dentex*	Increased mortality [137]

Table 3.1 (Continued)

Major species – EFA	Reported EFA deficiency signs and health impacts[a]
Turbot *Scophthalmus maximus*	Altered gill structure, disappearance of chloride cells, 'sloughing off' of epithelium along the primary & secondary filaments, accumulation of cellular material in inter-lamellar spaces [138]
White Fish *Coregonus lavaretus maraena*	Swollen pale livers with histological alterations [139]
White shrimp *Litopenaeus vannamei*	Reduced antioxidant enzyme activity [140]
Yellowtail *Seriola quinqueradiata*	Increased larval mortality (hydrops – floating bodies), disrupted schooling behavior [141]

Note: [a]Major reported deficiency signs and health impacts in addition to reduced growth and feed efficiency.

3.2 Oxidization of dietary lipids

Dietary lipids (including phospholipids, EFA, cholesterol, fat-soluble vitamins, and carotenoids) are highly prone to oxidative damage and degradation on prolonged storage; degradation usually being by autooxidation and combing with oxygen via free radical reactions and the subsequent formation of toxic oxidation products, including lipid hydroperoxides and secondary breakdown products [144–145, 683]. Table 3.2 shows the reported deleterious effects of feeding oxidized lipids to fish and crustaceans.

It is important to mention here, that the deleterious effects of oxidized lipids are twofold, namely (1) the direct toxic effects of the primary and secondary lipid degradation products (hydroperoxides, conjugated dienes/trienes, carbonyls, aldehydes, volatiles, malondialdehyde, oxysterols, etc.), and (2) the loss of essential nutrients through the oxidation process, such as the highly reactive long-chain polyunsaturated fatty acids (EPA, DHA) and tissue antioxidants, including α-tocopherol (vitamin E), vitamin C, carotenoids and antioxidant enzymes (superoxide dismutase, catalase and glutathione peroxidase) [145].

Notwithstanding the above, under practical farming conditions, the potential deleterious effects of oxidized or rancid dietary lipids can be reduced and/or eliminated through the increased use of

Table 3.2 Reported deleterious effects of feeding oxidized lipids and diets to fish and shrimp.

Major species	Reported pathology and health impacts
Arctic charr *Salvelinus alpinus*	Reduced growth & higher hepatosomatic index [146]
Atlantic cod *Gadus morhua*	Increased erythrocyte osmotic fragility [147], decreased liver alpha-tocopherol & increased oxidative stress [148]
Atlantic halibut *Hippoglossus hippoglossus*	Increased skeletal deformity: scoliosis & lordosis [149], reduced growth [150], reduced plasma glucose levels [151]
Atlantic salmon *Salmo salar*	Reduced oxidative stress [152], lipoid liver degeneration (LLD) & accumulation of ceroid in hepatocytes [153]
Black sea bream *Acanthopagrus schlegeli*	Reduced growth & survival rate, increased hepatosomatic index [154]
Channel catfish *Ictalurus punctatus*	Poor growth, feed efficiency & survival; exudative diathesis, muscular dystrophy & depigmentation; fatty liver & anemia [155], reduced lipid metabolism & fatty acid synthesis [156], reduced growth, liver & kidney histolgical alterations, increased lipid droplet accumulation in the hepatocytes, mitochondrial vacuolation in the renal tubules [157]
Chinese longsnout catfish *Leiocassis longirostris*	Lighter tail skin coloration [158]
Coho salmon *Oncorhynchus kisutch*	Reduced growth & feed efficiency [159]
Common carp *Cyprinus carpio*	Reduced growth, high mortality & muscular dystrophy or 'Sekoke' disease [160], muscular dystrophy [161], damaged gut mucosa [162]
European seabass *Dicentrachus labrax*	Increased erythrocyte fragility & innate immune response [166]
Grass carp *Ctenopharyngodon idella*	Severe intestinal mucosa damage [163], fatty liver, bile acid accumulation, increased oxidative stress & decreased antioxidant capacity of hepatopancreas [164], reduced intestine bile acid content, enhanced number of goblet cells, enlarged microvilli & gap between tight junction, hyperplasia & edema of villi, leading to damaged intestinal epithelial tight junction & destruction of the intestinal mucosal epithelium, and increased intestinal oxidative stress [165]

Table 3.2 Reported deleterious effects of feeding oxidized lipids and diets to fish and shrimp.

Major species	Reported pathology and health impacts
Hybrid catfish *Clarias macrocephalus* × *C. gariepinus*	Jaundice disease, high mortality, yellow skin & gill pigmentation, enlarged spleen, kidney & gall bladder, pale yellow spleen, kidney, liver & body fat, with yellow ascitic fluid in their abdomen; histologically heavy deposits of hemosiderin and ceroid in the spleen, kidney and liver; decreased hematocrit, RBC count, hemoglobin content, hemolytic anemia associated with lipoid liver degeneration – these symptoms were believed to be due to the feeding of rancid chicken viscera [167]
Largemouth bass *Micropterus salmoide*	Reduced growth & feed efficiency, increased oxidative damage, decreased hepatosomatic index [168], increased oxidative stress, histological alterations of liver with hepatocytes with lipid vacuoles and nuclear migration [169], severe oxidative damage & loss of reducing capacity [170]
Nile tilapia *Oreochromis niloticus*	Marked congestion, with some hemorrhage, in dermal vessels around snout and at bases of pectoral/dorsal fins, lordosis, exophthalmia, abdominal swelling, cataract, orbital collapse, darkening of liver, marked distension of bile duct, steatitis of all abdominal fat bearing tissue, deposits of intracellular ceroid in liver, spleen, kidney and choroid, increased mortality [189]
Rainbow trout *Oncorynchus mykiss*	Reduced growth & increased oxidative stress [171], lipoid liver degeneration due to feeds containing rancid lipids [172], reduced growth, microcytic anemia & liver lipoid degeneration [173], reduced growth & feed efficiency, severe muscle damage, increased mortality & increased erythrocyte fragility [174], reduced red blood cell numbers, hemoglobin content, hematocrit & increased hemolysis [175], increased mortality [176], rainbow trout fry syndrome (RTFS) – dystrophic changes in the liver, kidney and muscle [177], reduced growth & increased oxidative stress [178],
Red sea bream *Pagrus major*	Increased oxidative stress [179], reduced growth & increased oxidative stress [180]

Rohu *Labeo rohita*	Reduced growth & increased oxidative stress [181]
Senegalese sole *Solea senegalensis*	Reduced bone mineralization & skeletogenesis [182]
Siberian sturgeon *Acipenser baeri*	Reduced growth, survival, increased oxidative stress & larval deformity in the absence of Vitamin A [183]
Turbot *Scophthalmus maximus*	Muscular dystrophy [184], reduced growth [150], reduced disease resistance [185]
White shrimp *Litopenaeus vannamei*	Reduced growth, increased hepatosomatic index & oxidative stress [186]
Yellowtail *Seriola quinqueradiata*	Reduced growth & hematocrit, high mortality [187]
Yellow catfish *Pelteobagrus fulvidraco*	Reduced intestinal health & increased goblet cells in foregut [188]

dietary α-tocopherol [147–148, 150, 154–155, 166, 168, 173–176, 180–181], and to a lesser extent with supplemental vitamin C [179], vitamin A [183], selenium [168], vitamin premix [170], and glutathione [187]. In addition to the use of natural antioxidants, it is also common practice for feed ingredient suppliers and feed manufacturers to use synthetic antioxidants to prevent lipid peroxidation, including ethoxyquin, butylated hydroxytoluene (BHT), butylated hydroxyanisole (BHA), and propyl gallate (PG) [190–191]. Although the efficacy of synthetic antioxidants is generally less than α-tocopherol, their ready availability and lower cost has meant that most lipid-rich feed ingredient sources (fishmeal, fish oil, terrestrial animal by-product meals and fats) and finished aquaculture feeds are usually stabilized with synthetic antioxidants, and in particular ethoxyquin [110, 190–191]. However, due to increasing human food safety concerns regarding the possible carry-over effects of synthetic antioxidant residues into fish and shrimp meat [192], there is a growing trend toward the reduced use of synthetic antioxidants and the identification and use of more environmentally friendly natural antioxidant sources [193–194].

Other reported factors known to promote dietary lipid oxidation include (1) direct exposure of feed containing lipids to light (through photo-oxidation), (2) the presence of dietary pro-oxidants, such as inorganic trace elements (e.g. iron and copper), and (3) the presence of feed ingredients containing lipoxygenase enzymes, such as lipase within rice bran [195–196, 683].

4

Disorders in mineral nutrition

In common with most farmed animal species, fish and shrimp have a physiological requirement for over 20 minerals, including: (1) the macro-minerals: calcium (Ca), phosphorus (P), potassium (K), sodium (Na), sulphur (S), magnesium (Mg), chlorine (Cl) [110, 197–200]; (2) the trace minerals iron (Fe), zinc (Zn), copper (Cu), manganese (Mn), cobalt (Co), iodine (I), selenium (Se), chromium (Cr), molybdenum (Mo) [110, 197–200]; and (3) possibly the newer trace elements aluminium (Al), arsenic (As), boron (B), bromine (Br), flourine (F), germanium (Ge), lead (Pb), lithium (Li), nickel (Ni), rubidium (Rb), silicon (Si), tin (Sn), and vanadium (V) [201].

In marked contrast to terrestrial farm animals, fish and shrimp live immersed in an aquatic environment, and as such have the ability (depending upon the species and culture environment) to satisfy all and/or part of their dietary mineral requirement through mineral absorption across the gills, body surface, and/or through water ingestion [110, 197–200].

Tables 4.1–4.4 show the major reported deficiency signs and health impacts of fish and shrimp fed diets deficient in macro-minerals, including calcium (Table 4.1), phosphorus (Table 4.2), potassium (Table 4.3), and magnesium (Table 4.4). Moreover, Tables 4.5 to 4.10 show the major reported deficiency signs and health impacts of fish and shrimp fed diets deficient in trace minerals, including iron (Table 4.5), zinc (Table 4.6), copper (Table 4.7), manganese (Table 4.8), selenium (Table 4.9), and chromium, iodine, and cobalt (Table 4.10).

Table 4.1 Reported calcium (Ca) deficiency signs and health impacts on fish and shrimp.

CALCIUM (Ca)	Reported deficiency signs & health impacts
Atlantic salmon *Salmo salar*	Reduced bone mineralization when dietary phosphorus was below requirement level [202]
Blue tilapia *Oreochromis aureus*	Reduced growth & bone mineralization in low-calcium water [203], reduced growth & feed efficiency [204], reduced growth, feed efficiency & bone mineralization [205]
Channel catfish *Ictalurus punctatus*	Reduced growth & feed efficiency in calcium-free water [206–207],
Grass carp *Ctenopharyngodon idella*	Reduced growth, feed efficiency & bone mineralization [208, 216]
Tiger puffer *Takifugu rubripes*	
Hybrid tilapia *Oreochromis niloticus* × *O. aureus*	Reduced growth & bone/scale mineralization [209]
Indian major carp *Labeo rohita*	Reduced growth, feed efficiency, protein efficiency ratio, and whole-body, vertebrae & scale Ca content [308]
Japanese flounder *Paralichthys olivaceus*	Reduced growth [210, 214]
Red lip mullet *Liza haematocheila*	
Nile tilapia *Oreochromis niloticus*	Increased body deformity (short body) – closed recirculating aquaculture systems [211]
Grouper *Epinephelus coioides*	Reduced feed efficiency & scale mineralization [212]
Rainbow trout *Oncorynchus mykiss*	Delayed ossification of bone endochondral structures [213]
Red sea bream *Pagrus major*	Reduced bone K and Mn content [215]
White shrimp *Litopenaeus vannamei*	Reduced growth, feed efficiency & survival [217]

Notwithstanding the above reported mineral deficiency signs, it should be stated that the majority of these studies were performed with animals fed semi-purified experimental diets (so as to elicit a deficiency response [110, 197–199]), and as such may not be applicable to animals fed practical diets; animal and marine feed ingredient sources usually being a richer source of essential minerals and trace elements compared with plant feed ingredient sources, in which many minerals and trace elements are usually less biologically available [99, 110]. This difference between animal and plant

Table 4.2 Reported phosphorus (P) deficiency signs and health impacts on fish and shrimp.

PHOSPHORUS	Reported deficiency signs & health impacts
Atlantic salmon *Salmo salar*	Reduced growth, feed efficiency & bone mineralization [218], reduced whole-body P content [219], bone deformities and low Ca and P content of vertebrae [202]
Black seabream *Sparus macrocephalus* Chinese sucker *Myxocyprinus asiaticus* Grass carp *Ctenopharyngodon idella*	Reduced growth, feed efficiency, protein efficiency ratio & reduced bone, skin & scale mineralization [220, 224, 231]
Blue tilapia *Oreochromis aureus* Crucian carp *Carassius auratus* Red lip mullet *Liza haematocheila*	Reduced growth, feed efficiency & bone mineralization [204, 226, 243]
Catla *Catla catla*	Reduced growth & body P content [221]
Common carp *Cyprinus carpio*	Reduced growth, feed efficiency, bone mineralization, body ash and P, Ca & Mg content [222]
Channel catfish *Ictalurus punctatus* Chum salmon *Oncorhynchus keta* Japanese seabass *Lateolabrax japonicus* Milkfish *Chanos chanos* Mozambique tilapia *Oreochromis mossambicus* Siberian sturgeon *Acipenser baerii* Yellow catfish *Pelteobagrus fulvidraco*	Reduced growth & bone mineralization [223, 225, 233, 235–236, 244, 250]
European sea bass *Dicentrarchus labrax*	Reduced growth, feed efficiency, protein efficiency ratio & increased mortality [227]
Gibel carp *Carassius auratus gibelio var. CASIII*	Reduced growth, feed efficiency, ash & fish P content [228–229]
Gilthead sea bream *Sparus aurata*	Reduced growth [230]
Haddock *Melanogrammus aeglefinus*	Poor growth, loss of appetite, poor bone mineralization, deformed vertebrae and an increase in body lipid content [234]
Indian major carp *Labeo rohita*	Reduced growth, feed efficiency, protein efficiency ratio, alkaline phosphatase activity & P utilization efficiency [232]
Mrigal *Cirrhinus mrigala*	Reduced growth & feed efficiency [237]
Nile tilapia *Oreochromis niloticus* White shrimp *Litopenaeus vannamei*	Reduced growth, feed efficiency & survival [238, 252]
Olive Flounder *Paralichfhys olivaceus*	Reduced growth, feed efficiency & protein efficiency ratio [239]
Rainbow trout *Oncorynchus mykiss*	Reduced growth & body P content [240], reduced growth, feed efficiency, bone ash, and whole-body ash contents [241], reduced growth & bone mineralization [242]

Table 4.2 (Continued)

PHOSPHORUS	Reported deficiency signs & health impacts
Stinging catfish *Heteropneustes fossilis*	Reduced growth, feed efficiency, hematology (hemoglobin, hematocrit) & antioxidant status [248]
Striped bass *Morone saxatilis*	Reduced growth, feed efficiency & reduced scale, vertebral and dorsal fin mineralization, scoliosis [245]
Sunshine bass *Morone chrysops*	Reduced growth & bone mineralization, tetany [247]
Tambaqui *Colossoma macropomum*	Reduced bone mineralization and hematology [248]
Walking catfish *Clarias leather*	Reduced growth, protein efficiency ratio, body and vertebrae P content [249]
Yellow croaker *Pseudosciaena crocea*	Reduced growth, bone mineralization & body P content [251, 253]

Table 4.3 Reported potassium (K) deficiency signs and health impacts on fish and shrimp.

POTASSIUM (K)	Reported deficiency signs & health impacts
Chinook salmon *Oncorhynchus tschawytscha*	Anorexia, convulsions, tetany & death [254]
Fleshy prawn *Penaeus chinensis*	Reduced growth [260]
Giant tiger prawn *Penaeus monodon*	Reduced growth [261]
Grass carp *Ctenopharyngodon idella*	Reduced growth & antioxidant enzyme level [255–257]
Indian major carp *Labeo rohita*	Reduced growth, feed efficiency, protein efficiency ratio & antioxidant enzyme levels [258]
Kuruma prawn *Penaeus japonicus*	Reduced growth & protein efficiency ratio [259]

Table 4.4 Reported magnesium (Mg) deficiency signs and health impacts on fish and shrimp.

MAGNESIUM (Mg)	Reported deficiency signs & health impacts
Atlantic salmon *Salmo salar*	Reduced whole-body, serum and vertebrae Mg content [262]
Common carp *Cyprinus carpio*	Loss of appetite, poor growth, high mortality, sluggishness & convulsions, reduced vertebrae Mg content [263]

Table 4.4 (Continued)

MAGNESIUM (Mg)	Reported deficiency signs & health impacts
Grass carp *Ctenopharyngodon idella*	Reduced growth, feed efficiency & decreased whole-body, vertebrae, scales and plasma Mg content [264], reduced growth [265]
Hybrid tilapia *Oreochromis niloticus* × *O. aureus*	Reduced growth [266]
Kuruma prawn *Penaeus japonicus*	Reduced growth [261]
Rainbow trout *Oncorhynchus mykiss*	Reduced growth & electrolyte balance [267], reduced growth, appetite & movement, vertebral curvature & histological alterations of muscle, pyrolic caeca, and gill filaments [268]
Whiteleg shrimp *Litopenaeus vannamei*	Reduced growth [269]

Table 4.5 Reported iron (Fe) deficiency signs and health impacts on fish and shrimp.

Iron (Fe)	Reported deficiency signs and health impacts
Atlantic salmon *Salmo salar*	Reduced hematological values, liver and whole-body Fe content [270]
Channel catfish *Ictalurus punctatus*	Reduced growth, hemoglobin, hematocrit, plasma iron, transferrin saturation and erythrocyte count values [271]
Red sea bream *Chrysophrys major*	Reduced mean corpuscular constants of blood and Fe content [272]

Table 4.6 Reported zinc (Zn) deficiency signs and health impacts on fish and shrimp.

Zinc (Zn)	Reported deficiency signs and health impacts
Channel catfish *Ictalurus punctatus*	Reduced serum and bone Zn content [273], reduced growth [274–275]
Common carp *Cyprinus carpio*	Reduced appetite, growth & increased mortality, skin & fin erosion [276]
Grass carp *Ctenopharyngodon idella*	Reduced growth & whole-body, scale, vertebrae and plasma Zn content [277]
Grass shrimp *Penaeus monodon*	Reduced growth, whole-body Zn content & non-specific immune response [278]
Malabar grouper *Epinephelus malaricus*	Reduced growth & serum, muscle, vertebrae & scale Zn content [279]

Table 4.6 (Continued)

Zinc (Zn)	Reported deficiency signs and health impacts
Nile tilapia *Oreochromis niloticus*	Reduced growth, feed efficiency & bone Zn content [280], reduced growth & decreased liver, muscle, bone, and scale Zn content, reduced serum antioxidant enzymes, and altered liver histological signs (considerable nuclear migration) [281]
Orange-spotted grouper *Epinephelus coioides*	Reduced growth, feed efficiency, antioxidant enzyme activity & reduced whole-body and vertebrae Zn content [282]
Rainbow trout *Oncorynchus mykiss*	Reduced growth, high mortality, cataract, fin & skin erosion, reduced body Zn content [283]
Red drum *Sciaenops ocellatus*	Reduced bone & scale Zn content [284]
Siberian sturgeon *Acipenser baerii*	Reduced growth, feed efficiency, erythrocyte & leucocyte count, hemoglobin & hematocrit value [285]
Yellow catfish *Pelteobagrus fulvidraco*	Reduced growth & feed efficiency, reduced whole-body Zn content, reduced lipase & antioxidant (SOD) enzyme level [286]

Table 4.7 Reported copper (Cu) deficiency signs and health impacts on fish and shrimp.

Copper (Cu)	Reported deficiency signs and health impacts
Channel catfish *Ictalurus punctatus*	Reduced liver antioxidant enzyme level [287]
Cobia *Rachycentron canadum*	Reduced growth, feed efficiency, survival & whole-body and vertebrae Cu content [288]
Common carp *Cyprinus carpio*	Reduced growth in carp& tissue Cu content [289]
Indian catfish *Heteropneustes fossilis*	Reduced growth, feed efficiency & liver antioxidant enzyme activity [290]
Grass shrimp *Penaeus monodon*	Reduced growth, feed efficiency, protein efficiency, whole-body Cu content & non-specific immune response [291]
Malabar grouper *Epinephelus malaricus*	Reduced growth & liver antioxidant enzyme activity [292]
Orange-spotted grouper *Epinephelus coioides*	Reduced growth, feed efficiency & non-specific enzymes immunity [293]
Russian sturgeon *Acipenser gueldenstaedtii*	Reduced growth, feed efficiency & liver antioxidant activity [294]
Snakehead *Channa punctatus*	Reduced growth, feed efficiency, protein efficiency ratio, hemoglobin, hematocrit, RCBs & whole-body Cu content [295]

Table 4.7 (Continued)

Copper (Cu)	Reported deficiency signs and health impacts
White shrimp *Penaeus indicus*	Reduced growth, survival & whole-body Cu content [296]
Whiteleg shrimp *Litopenaeus vannamei*	Reduced growth, enlarged hearts & reduced hemolymph, carapace, and hepatopancreas Cu content [297]
Yellow catfish *Pelteobagrus fulvidraco*	Reduced growth & whole-body and liver Cu content [298]
Yellow croaker *Larimichthys croceus*	Reduced growth, total antioxidant activity & whole-body and vertebrae Cu content [299]

Table 4.8 Reported manganese (Mn) deficiency signs and health impacts on fish and shrimp.

Manganese (Mn)	Reported deficiency signs and health impacts
Atlantic salmon *Salmo salar*	Reduced whole-body Mn content & liver oxidation enzyme activity [300]
Cobia *Rachycentron canadum*	Decreased growth, feed efficiency, survival, whole-body, vertebrae & liver Mn content & decreased liver antioxidant enzyme activity [301]
Common carp *Cyprinus carpio*	Reduced growth [289]
Grass carp *Ctenopharyngodon idella*	Reduced growth, feed efficiency & whole-body, vertebrae and skin Mn content [302]
Hybrid grouper *Epinephelus lanceolatus* × *E. fuscoguttatus*	Reduced growth, feed efficiency, protein efficiency ratio, whole-body, liver and vertebrae Mn content, reduced antioxidant enzyme activity, and increased oxidative stress [303]
Hybrid tilapia *Oreochromis niloticus* × *O. aureus*	Reduced liver, bone and muscle Mn content & reduced liver antioxidant enzyme activity [305]
Rainbow trout *Oncorynchus mykiss*	Reduced growth, abnormal tail growth, short-body dwarfism [289]
Whiteleg shrimp *Litopenaeus vannamei*	Reduced growth [305]
Yellow catfish *Pelteobagrus fulvidraco*	Reduced growth & feed efficiency, decreased whole-body Mn content & decreased liver antioxidant activity [306]
Yellow croaker *Larimichthys croceus*	Reduced growth, feed efficiency, whole-body, liver and vertebrae Mn content, reduced antioxidant enzyme activity, and increased oxidative stress [307]

Table 4.9 Reported selenium (Se) deficiency signs and health impacts on fish and shrimp fed Selenium (Se) deficient diets.

Selenium (Se)	Reported deficiency signs and health impacts
Atlantic salmon *Salmo salar*	Reduced plasma antioxidant (Glutathione Peroxidase – GPx) enzyme activity [309]
Channel catfish *Ictalurus punctatus*	Reduced growth & liver and plasma antioxidant enzyme activity [310]
Cobia *Rachycentron canadum*	Reduced growth, feed efficiency, survival & whole-body and vertebrae Se content, and reduced antioxidant enzyme activity [311]
Gibel carp *Carassius auratus gibelio var. CASIII*	Reduced growth & antioxidant capability [312]
Japanese seabass *Lateolabrax japonicus*	Reduced growth [313, 318]
Rainbow trout *Salmo gairdneri*	
Malabar grouper *Epinephelus malabaricus*	Reduced growth [314], reduced growth, feed efficiency & liver antioxidant enzyme activity [315]
Nile tilapia *Oreochromis niloticus*	Reduced growth & antioxidant capability [316], reduced growth, feed efficiency, survival & antioxidant enzyme activity [317]
Yellow croaker *Larimichthys croceus*	Reduced growth & whole-body and vertebrae Se content, reduced total antioxidant activity [319]

Table 4.10 Reported chromium (Cr), iodine (I), and cobalt (Co) deficiency signs and health impacts on fish and shrimp.

Chromium (Cr)	Reported deficiency signs and health impacts
Common carp *Cyprinus carpio*	Reduced growth & carbohydrate utilization [320]
Grass carp *Ctenopharyngodon idellus*	Reduced growth, feed efficiency, protein efficiency ratio, protein retention, and altered carbohydrate & lipid metabolism [321]
Indian major carp *Labeo rohita*	Reduced growth, feed efficiency, protein efficiency ratio & apparent net protein utilization [322]
Nile tilapia *Oreochromis niloticus*	Reduced growth [323]
Rainbow trout *Oncorhynchus mykiss*	Reduced immune response [324], reduced growth & feed efficiency [325]
Iodine (I)	
Fleshy prawn *Penaeus chinensis*	Reduced growth & survival [260]
Cobalt (Co)	
Malabar grouper *Epinephelus malabaricus*	Reduced growth [326]

feed ingredient sources is particularly important for marine protein (that is, fishmeal) dependent cultured species, and the necessity to reduce their dependency on these finite feed resources with alternative terrestrial land animal and plant-based feed ingredient sources [197–198].

As a general rule, animals reared under nutrient-poor freshwater environments are more likely to require dietary mineral and trace element supplementation than animals reared under marine culture conditions, and as such feeds should be formulated using a dietary mineral and trace element premix. Moreover, in view of the pro-oxidant properties of many inorganic mineral salts, it would be advisable to use minerals chelated to organic molecules (such as specific amino acids or microorganisms) so as to maximize biological availability, and at the same time limit unwanted biological interactions and nutrient leaching in the case of soluble inorganic mineral salts [198, 274].

As with other dietary essential nutrients, many minerals and trace elements have important functional properties, and as such these minerals may require higher dietary fortification for maximum health and immune response, and in particular the immune stimulating trace elements zinc, copper, manganese, selenium, and chromium [278, 291, 324, 327].

However, it is also important to mention here that most minerals and trace minerals may exert a negative effect on fish/shrimp growth and health if supplied in excess of dietary requirement, either due to direct toxicological effects or due to negative interactions with other minerals. For example, an excess of dietary calcium (calcium carbonate or limestone often being used as an inexpensive filler in aquafeeds) has been reported to exert a negative effect on the bone mineralization and the biological availability of other essential minerals, including P, Mg, K, Zn, Mn, and Fe [210, 212, 214–216, 328]. As a general guideline, the dietary Ca : P ratio should be kept at about <2 : 1 [110, 329].

5

Disorders in vitamin nutrition

By far the commonest nutritional disorder reported in farmed fish and shrimp arise from the use of feeds deficient in one or more essential dietary vitamins [7, 110]. The vitamins represent a diverse group of organic compounds required by all animal species for the maintenance of the life process, and are commonly divided into two major groups according to their solubility, namely:

- the fat-soluble vitamins: vitamin A (retinol), vitamin D (D_3 – cholecalciferol), vitamin E (tocopherol), and vitamin K (K_3 – manadione);
- the water-soluble vitamins: B-complex vitamins – vitamin B_1 (thiamine), vitamin B_2 (riboflavin), vitamin B_6 (pyridoxine), pantothenic acid, nicotinic acid (niacin), biotin, folic acid, vitamin B_{12} (cyanocobalamin), and water-soluble macro-vitamins – inositol (myo-inositol), choline, and vitamin C (ascorbic acid).

Tables 5.1, 5.2, and 5.3 show the reported deficiency signs for the fat-soluble vitamins, water-soluble B-complex vitamins, and water-soluble macro-vitamins, respectively.

Although dietary vitamin deficiencies can be readily resolved through the purchase and use of specific dietary vitamins and/or the use of commercial multi-vitamin premixes (usually formulated to published established guidelines [110]), to date the majority of published data available has been based on the use of semi-purified diets

Table 5.1 Reported fat-soluble vitamin deficiency signs and health impacts on fish and shrimp.

Vitamin	Reported deficiency signs and health impacts
Vitamin A (retinol)	
Cherry salmon *Oncorhynchus masou*	Reduced growth, pale body color, red fins, truncated snout, fish survival & severe anemia, poorly formed bony tissue in the skull and vertebral protuberance, degeneration of skeletal and cardiac muscle fibers, and thinning of the epidermis with a decrease of mucous cells [330]
Fleshy prawn *Penaeus chinensis*	Reduced growth & visual organ development [331], reduced growth, metamorphosis & larval survival [332]
Grass shrimp *Penaeus mondon*	Reduced growth & survival [333]
Greasy grouper *Epinephelus tauvina*	Reduced growth, feed efficiency, survival, skin hemorrhage at fin base & erosion of caudal peduncle [334]
Japanese flounder *Paralichthys olivaceus*	Reduced growth [335]
Jian carp *Cyprinus carpio* var. *Jian*	Reduced growth, feed efficiency, survival, blood erythrocyte & leukocyte counts, and immune response [336], anorexia, reduced growth, loss of body coloration, exophthalmia, sluggish movement, hemorrhage in the skin, fins & mouth, spinal deformities, increased mortality, reduced intestinal fold height, reduced erythrocyte & leucocyte count, and decreased non-specific immune response [337]
Wuchang bream *Megalobrama amblycephala*	Reduced growth, protein efficiency ratio, red & white blood cell count & hemoglobin content, skeletal deformity, eye lesions, hemorrhages of fins and eyes, and reduced immunity [338]
Vitamin D_3 (cholecalciferol)	
Channel catfish *Ictalurus punctatus*	Poor growth, lowered body calcium, phosphorus & ash levels [343]
Fleshy prawn *Penaeus chinensis*	Reduced growth, phosphorus & calcium absorption, & shell hardening [339]
General – Fish	Reduced skeletal ossification & increased skeletal deformities [340]
Hybrid tilapia *Oreochromis niloticus* × *O. aureus*	Reduced growth & feed efficiency, low hemoglobin & hepatosomatic index [341]
Rainbow trout *Oncorynchus mykiss*	Poor growth, elevated liver lipid content, impaired calcium homeostasis, manifested by tetany of white skeletal muscles [344], lordosis-like droopy tail syndrome [345]
Wuchang bream *Megalobrama amblycephala*	Reduced growth & feed efficiency, decreased plasma Ca content [342]

Table 5.1 (Continued)

Vitamin	Reported deficiency signs and health impacts
Vitamin E (tocopherol)	
Atlantic salmon *Salmo salar*	Reduced growth & evidence of skeletal muscle degeneration [346], impaired immune response [347], anemia, pale gills, anisocytosis, poikilocytosis, elevated plasma protein, exudative diathesis, dermal depigmentation, in vitro ascorbic acid-stimulated peroxidation in hepatic microsomes, yellow-orange liver color, yellow-brown intestinal contents, enlarged gall bladder distended with dark green bile, muscular dystrophy, and a response to handling characterized by a transitory fainting with interruption in swimming [348]
Barramundi *Lates calcarifer*	Lateral muscular myopathy [349]
Beluga sturgeon *Huso huso*	Reduced growth & feed efficiency [350]
Blue tilapia *Oreochromis aureus*	Reduced growth, feed efficiency, skin and fin hemorrhages, anorexia, impaired erythropoiesis, muscle degeneration, ceroid depositions in liver & spleen, and lack of skin color [351]
Channel catfish *Ictalurus punctatus*	Extreme atrophy and necrosis of white muscle fibers, sclerotic glomeruli in the posterior kidney & ceroid-laden macrophages in intrahepatic blood vessels [410]
Cherry/amago salmon *Oncorhynchus masou*	Reduced growth, anorexia, mild anemia, hepatic lesions, with periarteritis occasionally observed in the submucosa layer of intestine [352]
Common carp *Cyprinus carpio*	Increased mortality, 'thin back disease', lordosis and exudative diathesis with exophthalmus, scale erection and ascites, skeletal muscular dystrophy, swollen yellow/blue colored liver (the liver showing degeneration, necrosis with numerous yellow or brown-yellow ceroid-like dense bodies in liver cells), enlarged heart with hemorrhage in epicardium, swollen kidney with hyperemia and hemorrhage [353], reduced growth & apparent muscular dystrophy [412]
Fleshy prawn *Penaeus chinensis*	Reduced growth, reproductive performance & antioxidant capacity [367]
Grass carp *Ctenopharyngodon idella*	'Sekoke disease' characterized by a marked loss of flesh in the back [409]
Grass shrimp *Penaeus monodon*	Reduced growth, total hemocyte count & antioxidant (superoxide dismutase) activity [354]
Grouper *Epinehelus moara*	Reduced growth, feed efficiency, survival & immune response [355]

Table 5.1 (Continued)

Vitamin	Reported deficiency signs and health impacts
Hybrid striped bass *Morone chrysops* female × *M. saxatilis* male	Reduced growth, feed efficiency, hematocrit & dark coloration [356]
Indian major carp *Cirrhinus mrigala*	Reduced growth, feed efficiency & protein efficiency ratio, increased erythrocyte fragility [357]
Oriental river prawn *Macrobrachium nipponense*	Reduced growth & immune response [358]
Rainbow trout *Oncorhynchus mykiss*	Reduced growth, feed efficiency, hematocrit, erythrocyte fragility, severe muscle myopathy & increased mortality [359], increased mortality, muscular dystrophy & splenic hemosiderosis, decreased hematocrit, hemoglobin content & red blood cell count, increased spontaneous erythrocyte fragility [360], anorexia, reduced growth, severe convulsions, increased mortality [411]
Red drum *Sciaenops ocellatus*	Reduced growth & feed efficiency, heart edema, reduced respiratory burst of head kidney macrophages [361]
Rohu *Labeo rohita*	Reduced growth, feed efficiency & increased erythrocyte fragility [362]
Siberain sturgeon *Acipenser Baerii*	Reduced growth, feed efficiency, hemato-biochemical status & imminue response [363]
Spotted snakehead *Channa punctatus*	Reduced growth, feed efficiency, protein/energy retention efficiency, hematocrit & hemoglobin content [364]
Whiteleg shrimp *Litopenaeus vannamei*	Reduced growth & antioxidant status [365]
Yellowtail kingfish *Seriola lalandi*	Reduced growth & serum lysozyme activity, and evidence of myopathy in fish deficient in both Se and vitamin E [366]
Vitamin K_3 (menadione)	
Fleshy prawn *Penaeus chinensis*	Reduced growth, feed efficiency, protein efficiency ratio, lower prothrombin levels [368]
General – fish	Mortality, anemia, increased blood clotting time and histopathological changes in liver and gills [369]
Haddock *Melanogrammus aeglefinus*	Reduced bone mineralization & increased bone deformities [370]

Table 5.2 Reported water-soluble B-complex vitamin deficiency signs and health impacts on fish and shrimp.

Vitamin	Reported deficiency signs and health impacts
Vitamin B_1 (thiamine)	
Atlantic salmon *Salmo salar*	'Cayuga Syndrome' maternally transmitted & causing 100% mortality of larval landlocked Atlantic salmon, gross lesions including yolk-sac opacities, subcutaneous edema, vitelline hemorrhage or congestion, pericardial edema, retrobulbar edema, branchial congestion, foreshortened maxillae, hydrocephalus, and occasional caudal fin deformities [371]
Barramundi *Lates calacarifer*	Anorexia, reduced growth, feed efficiency & survival, darkened coloration, sluggish swimming behavior, and nervous disorders including sensitivity to shock, convulsions and rapid mortality from physical shock [381]
Channel catfish *Ictalurus punctatus*	Anorexia, extremely poor growth, dark coloration of the skin and increased mortality rates [384]
European eel *Anguilla anguilla*	Deteriorated erythrocyte system, including increased erythroblast and prerythrocyte counts, enlargement of nuclei, deformities of cells, cases of endotosis & amitosis and one case of hypochromia [385]
General – Fish	Reduced growth, feed efficiency & neurological disorders such as hyperirritability and increased mortality [110, 377], clouding of the cornea [383]
Grass carp *Ctenopharyngodon idella*	Decreased growth and immunity of intestine, head kidney, spleen and liver [378], reduced growth, feed efficiency, protein efficiency ratio & hepatosomatic index [380]
Indian major carp *Cirrhinus mrigala*	Reduced growth, protein retention, hemoglobin, hematocrit & red blood cell count [376]
Japanese eel *Anguilla japonica*	Reduced growth, hemorrhage & ataxia [372]
Kuruma prawn *Penaeus japonicus*	Reduced growth [375]
Lake Trout *Salvelinus namaycush*	Acute larval mortality – also known as early mortality syndrome [373–374]
Nile tilapia *Oreochromis niloticus*	Reduced growth [379]
Rainbow trout *Oncorhynchus mykiss*	Nervous system disorders, loss of equilibrium accompanied by whirling, melanotic appearance, inability to feed, progressive weakness, and paralysis [387], anorexia, darkening and ataxia, death rapidly following overt thiamin deficiency [389]
Turbot *Scophthalmus maximus*	Reduced growth, increased mortality [386]

Table 5.2 (Continued)

Vitamin	Reported deficiency signs and health impacts
Vitamin B_2 (riboflavin)	
Barramundi *Lates calacarifer*	Anorexia, reduced growth, feed efficiency, photophobia, sluggish behavior, dark coloration, cataract, hyperirritability, short-body dwarfism, high mortality [381]
Blue tilapia *Oreochromis aureus*	Anorexia, reduced growth, anemia, lethargy, fin erosion, loss of normal body color, short-body dwarfism & cataracts [390]
Brook trout *Salvelinus fontinalis*	Anemia [391]
Channel catfish *Ictalurus punctatus*	Anorexia, reduced growth & short-body dwarfism [392]
Common carp *Cyprinus carpio*	Anorexia, poor growth, high mortality, hemorrhage of skin & fins [393]
Grass carp *Ctenopharyngodon idella*	Anorexia, reduced growth & gill immunity, impaired gill antioxidant system & gill tight junction proteins [394]
Grouper *Epinephelus coioides*	Reduced liver superoxide dismutase and catalase activity, antioxidant capacity and susceptibility to lipid peroxidation [388]
Hybrid red tilapia *Oreochromis spp*	Anorexia, reduced growth & hematocrit, short-body dwarfism, lens cataracts, high mortality [382]
Rainbow trout *Oncoryhnchus mykiss*	Anorexia, reduced growth, lethargy, apparent muscular weakness, severe fin erosion (all fins), rapid opercular movements, light greenish-yellow skin coloration and high mortality rate [395], anorexia, poor growth, high mortality, lesion of the fins & cataract [393], cataract [396, 399]
Spotted snakehead *Channa punctatus*	Reduced growth, feed efficiency, protein efficiency ratio & reduced antioxidant capability [397]
Sunshine bass *Morone chrysops* × *Morone saxatilis*	Anorexia, reduced growth & feed efficiency, dark body color & cataracts [398]
Vitamin B_6 (pyridoxine)	
Atlantic salmon *Salmo salar*	Increased mortality; behavioral changes; degenerative changes in kidneys, ovaries, and liver; a paucity of thyroid colloid; and hyperplasia of renal hematopoietic tissue [400]
Blunt snout bream *Megalobrama amblycephala*	Reduced growth, feed efficiency, survival, protein efficiency ratio, and nitrogen retention efficiency [401]
Gilthead seabream *Sparus aurata*	Reduced growth & feed efficiency, high mortality, hyperirritability coupled with erratic swimming behavior and degenerative changes in peripheral nerves [402]

Table 5.2 (Continued)

Vitamin	Reported deficiency signs and health impacts
Grass shrimp *Penaeus monodon*	Reduced growth & hepatopancreatic GOT activity [414]
Hybrid red tilapia *Oreochromis spp*	Anorexia, reduced growth & feed efficiency, abnormal neurological signs and convulsions followed by muscle spasms and lethargy, reduced hematocrit, severe caudal fin erosion and mouth lesions, high mortality [403]
Indian catfish *Heteropneustes fossilis*	Reduced growth, feed efficiency & protein efficiency ratio, anorexia, lethargy, pale body color, tetany, convulsions and hemorrhages in intestine and kidney, pale livers, spleens and gills [404]
Indian major carp *Cirrhinus mrigala*	Reduced growth, feed efficiency, protein retention, RNA/DNA ratio, hemoglobin, hematocrit & red blood cell count [407]
Jian carp *Cyprinus carpio* var. *Jian*	Reduced growth & intestine amylase activity [405]
Kuruma prawn *Penaeus japonicus*	Reduced growth, high mortality [375]
Orange-spotted grouper *Epinephelus coioides*	Anorexia, poor survival, convulsion, helically swimming and hyperirritability, reduced hemoglobin, and antioxidant capability [406]
Spotted snakehead *Channa punctatus*	Reduced growth & feed efficiency [408], nervous disorders such as ataxia, hyperirritability and muscular spasm followed by convulsions, fits, anorexia and paralysis in the posterior region [413]
Pantothenic acid	
Barramundi *Lates calacarifer*	Anorexia, reduced growth, feed efficiency; dark coloration, loss of equilibrium, abnormal swimming behavior; hemorrhaging operculum, lower jaw, isthmus & pelvic fin; eroded pelvic fin; clubbed gills & high mortality [381]
Blue tilapia *Oreochromis aureus*	Poor growth, hemorrhage, sluggishness, high mortality, anemia and severe hyperplasia of the epithelial cells of gill lamellae [415]
Brook trout *Salvelinus fontinalis*	Anemia, gill disease [416]
Channel catfish *Ictalurus punctatus*	Reduced growth, feed efficiency & gill lesions [417]
Grass carp *Ctenopharyngodon idella*	Reduced growth & feed efficiency, and immune capability [418]
Golden pompano *Trachinotus ovatus*	Reduced growth & feed efficiency, and immune capability [419]

Table 5.2 (Continued)

Vitamin	Reported deficiency signs and health impacts
Hybrid striped bass *Morone chrysops* × M. *saxatilis*	Reduced growth, hemorrhages, sluggishness, high mortality, anemia and severe hyperplasia of the epithelial cells of gill lamellae [420]
Lake trout *Salvelinus namaycush*	Anorexia, inanition, emaciation, gill abnormalities and high mortality [421]
Malabar grouper *Epinephelus malabaricus*	Reduced growth, anorexia, clubbed gills, anemia, sluggishness & high mortality [422]
Spotted snakehead *Channa punctatus*	Reduced hemoglobin, hematocrit, red blood cell count & anemia, reduced antioxidant capability [423]
Spotted rose snapper *Lutjanus guttatus*	Reduced growth, caudal fin erosion, desquamation of the skin, hemorrhages in the body and fins, lethargy and increased mortality [424]
Nicotinic acid (niacin)	
Blunt snout bream *Megalobrama amblycephala*	Reduced growth, feed efficiency, protein efficiency ratio, and increased viscerosomatic index, hepatosomatic index and mesenteric fat index [425]
Grass carp *Ctenopharyngodon Idella*	Reduced growth, feed efficiency, anorexia, digestive enzyme activity [426], reduced gill immunity & antioxidant capability, and gill tight junction proteins and related signaling molecules [427], reduced intestinal mucosal immune and intestinal physical function [428]
Grass shrimp *Penaeus monodon*	Reduced growth & feed efficiency [429]
Indian catfish *Heteropneustes fossilis*	Reduced growth, protein efficiency ratio & survival, reduced hematocrit, anemia, anorexia, lethargy and skin hemorrhage [430]
Nile tilapia *Oreochromis niloticus*	Reduced growth [431]
Rainbow trout *Oncorhynchus mykiss*	Reduced growth, feed efficiency & increased susceptibility to UV-induced skin damage [432]
Spotted snakehead *Channa punctatus*	Reduced growth & feed efficiency; reduced hemoglobin, hematocrit, and red blood cell count; difficult motion and skin hemorrhage, high mortality [433]
Folic acid	
Channel catfish *Ictalurus punctatus*	Reduced hematocrit [467]
Grass carp *Ctenopharyngodon idella*	Impaired intestinal immunity, antioxidant capacity and tight junctions [434], impaired fish gill health status via regulating gene expression of cytokines, tight junction proteins, antioxidant enzymes [435]
Grass shrimp *Penaeus monodon*	Reduced growth & feed efficiency [436]

Table 5.2 (Continued)

Vitamin	Reported deficiency signs and health impacts
Malabar grouper *Epinephelus malabaricus*	Reduced growth, feed efficiency & non-specific immune capability [437]
Nile tilapia *Oreochromis niloticus*	Reduced growth, red blood cell, hemoglobin & hematocrit [438]
Biotin	
Blunt snout bream *Megalobrama amblycephala*	Reduced growth, anorexia, protein efficiency ratio & survival rate [439]
Catla *Catla catla*	Reduced growth, feed efficiency & protein efficiency ratio; reduced hematocrit, hemoglobin and red blood cell count [440]
Hybrid tilapia *Oreochromis niloticus* × *O. aureus*	Reduced growth [444]
Indian catfish *Heteropneustes fossilis*	Anorexia, poor feed efficiency, convulsions, listlessness, dark skin color, tetanus and heavy mortality [441]
Japanese seabass *Lateolabrax japonicus*	Anorexia, poor growth, dark skin color, atrophy and high mortality [442]
Lake trout *Salvelinus namaycush*	Reduced growth, gill and liver lesions [421]
Nile tilapia *Oreochromis niloticus*	Reduced growth, lethargy, anorexia, circular swimming and convulsions, leading to death [443]
Rainbow trout *Oncorhynchus mykiss*	Reduced growth & feed efficiency [445]
Spotted snakehead *Channa punctatus*	Reduced growth, feed efficiency & protein efficiency ratio [446]
Walking catfish *Clarias batrachus*	Reduced growth, feed efficiency & protein efficiency ratio, anorexia, dark skin color and convulsions, higher mortality [447]
Vitamin B_{12}	
Blunt snout bream *Megalobrama amblycephala*	Reduced growth, feed efficiency & intestinal protease, amylase and lipase activity, and intestinal digestive capability [448]
Grass shrimp *Penaeus monodon*	Reduced growth [449]

Table 5.3 Reported water-soluble macro-vitamin deficiency signs and health impacts on fish and shrimp.

Macro-vitamin	Reported deficiency signs and health impacts
Choline	
Atlantic salmon *Salmo salar*	Lipid malabsorption syndrome characterized by pale and foamy appearance of the hypervacuolated enterocytes of the pyloric caeca (the result of lipid accumulation); reduced lipid transport, phospholipid synthesis & lipoprotein secretion capability [450–451]

Table 5.3 (Continued)

Macro-vitamin	Reported deficiency signs and health impacts
Blunt snout bream *Megalobrama amblycephala*	Reduced growth & increased liver lipid content [452]
Cobia *Rachycentron canadum*	Reduced growth, feed efficiency, survival, hepatosomatic index & muscle lipid content [453]
Fleshy prawn *Penaeus chinensis*	Reduced growth, survival & exuviation rate [454]
Gibel carp *Carassius auratus gibelio*	Reduced growth & increased carcass lipid content [455]
Hybrid tilapia *Oreochromis niloticus* × *O. aureus*	Reduced feed efficiency, survival & blood triglyceride, cholesterol and phospholipid concentrations [457]
Jian carp *Cyprinus carpio* var. Jian	Depression of intestinal mucosal immune system by inducing inflammation and dysfunction of the intestinal physical barrier, and regulating related signaling molecules of fish [456]
Olive flounder *Paralichthys olivaceus*	Reduced growth, feed efficiency & protein efficiency ratio [458]
Red drum *Sciaenops ocellatus*	Reduced growth, feed efficiency & liver lipid content [459]
Whiteleg shrimp *Litopenaeus vannamei*	Increased muscle lipid and phospholipid content [460]
White sturgeon *Acipenser transmontanus*	Reduced growth & extensive liver diffused fat vacuolation and fatty cyst formation [461]
Yellow perch *Perca flavescens*	Reduced growth, feed intake & high hepatic lipid content [462]
Inositol (myo-inositol)	
Fleshy prawn *Penaeus chinensis*	Reduced growth, survival & exuviation rate [454]
Gibel carp *Carassius auratus gibelio*	Reduced growth, feed efficiency, protein efficiency ratio & survival [463]
Grass shrimp *Penaeus monodon*	Reduced growth, feed efficiency & survival [464]
Hybrid sturgeon *Acipenser baerii* × *A. schrenckii*	Reduced growth & total antioxidant capacity [465]
Olive flounder *Paralichthys olivaceus*	Reduced growth & survival, abnormal lipid metabolism [466]
Vitamin C (ascorbic acid)	
African catfish *Clarias gariepinus*	Broken-skull disease, including hemorrhages along the spinal column and inflammation of dorsal skull bone followed by broken skull and death [468], anemia – reduced red blood cell levels, hemoglobin & hematocrit; leukopenia & increased susceptibility to pathogenic agents; hypertrophy, oedematous separation & telangiectasia of gill lamellae; shrunken glomeruli, sloughed Bowman's capsules & tubular dilation and distortion; shrunken hepatocytes and occurrence of yellowish ceroid pigments [470]

Table 5.3 (Continued)

Macro-vitamin	Reported deficiency signs and health impacts
Atlantic halibut *Hippoglossus hippoglossus* L.	Skeletal abnormalities, including scoliosis & lordosis primarily in the hemal region of the vertebral column [469]
Atlantic salmon *Salmo salar*	Lethargy, scoliosis, lordosis, broken back and anemia [471], reduced growth, deformed vertebrae, increased mortality, decreased mean volume of red blood cells [472], reduced growth, increased mortality, and reduced hydroxyproline content in skin and vertebrae [473], reduced immune capability [347]
Ayu *Plecoglossus altivelis*	Reduced growth & survival, gill deformation and hemorrhagic exophthalmia [474]
Channel catfish *Ictalurus punctatus*	Darker skin pigment, scoliosis, lordosis, and eroded fins [475], reduced growth, deformed spinal columns, external and internal hemorrhages, erosion of fins, dark skin color, reduced wound repair, distortion of gill filament cartilage [476]
Chinese sucker *Myxocyprinus asiaticus*	Reduced antioxidant capability [477]
Cobia *Rachycentron canadum*	Poor growth & higher mortality [478]
Coho salmon *Oncorhynchus kisutch*	Acute lordosis or scoliosis with resultant high mortality, reduced wound repair [479]
Giant river prawn *Macrobrachium rosenbergii*	Decreased molting ability & increased mortality [481]
Gilthead bream *Sparus aurata*	Anorexia, scale loss, depigmentation, internal and external hemorrhages, increased mortality, reduced wound healing capability, kidney granulomatosis [480]
Grass carp *Ctenopharyngodon idella*	Decrease fish gill immune function, disrupted physical barriers and immune barriers [482], reduced growth, anorexia, decrease fish head kidney and spleen immunity & oxidative capability [483]
Grass shrimp *Penaeus monodon*	Reduced growth [484], reduced growth, feed efficiency & survival [485], black spot, erratic swimming, loss of appetite, incomplete molting, soft-shell, sluggishness, flipped gill cover and high mortality [486],
Hybrid tilapia *Oreochromis niloticus* × *O. aureus*	Reduced growth, feed efficiency & protein efficiency ratio; low hematocrit and mortality [487], reduced growth and hepatic ascorbate concentration [488]
Hybrid striped bass *Morone chrysops* × *M. saxatilis*	Reduced growth, reduced plasma and liver ascorbic acid levels, and abnormalities in isthmus cartilage formation [489]

Table 5.3 (Continued)

Macro-vitamin	Reported deficiency signs and health impacts
Indian major carp *Labeo rohita*	Reduced growth, feed efficiency & protein efficiency ratio; reduced survival and behavioral abnormalities such as lethargic movements and poor feed intake [490]
Kuruma prawn *Penaeus japonicus*	Reduced growth & increased mortality, reduced resistance to stress [491]
Largemouth bass *Micropterus salmoides*	Poor growth, feed efficiency, survival & anorexia; lordosis, skin darkening, sluggish behavior, exophthalmia, reduced liver antioxidant capability [492]
Malabar grouper *Epinephelus malabaricus*	Reduced growth & non-specific immune capability [493]
Nile tilapia *Oreochromis niloticus*	Reduced growth & muscle collagen content; reduced whole-body lipid content; reduced serum alkaline phosphatase, aspartate aminotransferase and alanine aminotransferase [494]
Pacu *Piaractus mesopotamicus*	Reduced growth, hyperplasia, hypertrophy and dysplasia of the bone cartilage of gill filaments, anorexia and increased handling stress [495]
Rainbow trout *Oncorhynchus mykiss*	Reduced growth rate, hemorrhages and gill alterations as well as severe deformations and fractures of the vertebral column [360], distorted and twisted filamental cartilage, distension of eye scleral cartilage, acute spinal column damage [496], decreased hemoglobin content [497], lordosis, scoliosis, and exophthalmia [498], acute lordosis or scoliosis with resultant high mortality, reduced wound repair [479], exophthalmia, anorexia and lethargy, lordosis, and scoliosis [499]
Red drum *Sciaenops ocellatus*	Reduced growth, lordosis, scoliosis and loss of equilibrium [500]
Spotted rose snapper *Lutjanus guttatus*	Fin erosion, dark skin, desquamation and erratic swimming [501]
Spotted snakehead *Channa punctatus*	Hypo-functioning of thyroid gland [502], hyprochromic anemia, leukocytosis and decrease in neutrophils and lymphocytes [503], reduced growth, increased mortality, skeletal abnormalities (lordosis, scoliosis and degenerative changes in the thoracic vertebrae), increased liver cholesterol content [504]
Turbot *Scophthalmus maximus*	Tyrosinaemia and renal granulomatous disease [505]
Vundu catfish *Heterobranchus longifilis*	Reduced growth, lordosis, caudal fin deformity, skin erosion [506]

Table 5.3 (Continued)

Macro-vitamin	Reported deficiency signs and health impacts
Whiteleg shrimp *Litopenaeus vannamei*	Reduced survival [507]
Yellow perch *Perca flavescens*	Scoliosis, lens cataracts, anorexia, and hemorrhages [508]
Walleye *Stizostedion vitreum*	Retarded growth, increased mortality, skeletal deformities (primarily lordosis); twisted gill filaments & dislocation of vertebrae; vertebral lesions included focal hemorrhage, compressed spinal cord, and displacement of adjacent kidney and skeletal muscle; isthmus breakage; occasional hemorrhagic fins, skin, and eyes [509]

(so as to elicit a deficiency response) fed to juvenile animals reared under controlled indoor laboratory conditions; with specific vitamin requirement levels usually being ascertained singly, and based on optimum growth and feed efficiency, tissue vitamin storage level, and/or the prevention of overt deficiency signs as the requirement selection criteria.

However, while this published information is useful as an initial reference or guideline, under commercial feed manufacturing and farming conditions, these requirements may need to be modified or adjusted as follows.

- Commercial aquaculture feeds may contain feed ingredients that are a rich source of important vitamins, and as such the supplementation of specific dietary vitamins may be reduced or eliminated (that is, such as the use of soy lecithin as a supplementary source of dietary choline and inositol, the use of marine fish oils as a supplementary source of vitamin A and D, and the use of brewer's yeast as a supplementary source of B-vitamins [110, 510].
- Certain feed ingredients contain anti-vitamin factors that have a negative effect on vitamin stability and activity unless adequately processed, including (1) anti-thiamine factor (thiaminase) present within certain pelagic fish species and

fish silage, cottonseed, linseed, mung bean, and mustard seed [110, 377, 511], (2) anti-vitamin A factor present in soybeans, which destroys carotene and is not readily destroyed by heat, (3) anti-vitamin D factor present in soybeans and is destroyed by autoclaving, (4) anti-vitamin E factor present in kidney beans, soybeans, alfalfa and field pea and is destroyed by autoclaving, (5) anti-vitamin K factor present in sweet clover, (6) anti-niacin factor present in sorghum, (7) anti-pyridoxine factor present in linseed, which is destroyed by water extraction and autoclaving, (8) anti-vitamin B_{12} factor present in raw soybeans, and (9) anti-biotin factor (avidin) found in raw hens' eggs [512–514].

- Sensitivity of individual vitamins to destruction and loss depending upon external environmental factors (Table 5.4), such as vitamin losses due to feed processing through direct exposure to steam and heat during the pelleting and/or extrusion feed manufacturing process (Table 5.5 [515–516]), vitamin losses due to poor feed handling and storage conditions and subsequent exposure to variable ambient temperature, humidity, direct sunlight, and/or mold/insect infestation (Table 5.6 [517–518]), and vitamin losses due to nutrient

Table 5.4 Sensitivity of vitamins to external factors [518].

Vitamin	Heat	Oxygen	Humidity	Light
A	+	+	0	+
D_1	0	+	0	0
E	–	0	–	0
K_1	0	0	+	–
B_1	0	0	0	
B_2	–	–	0	0
B_6	+	–	0	0
B_{12}	0	0	0	0
Pantothenic acid	0	–	+	
Niacin	–	–	–	–
Biotin	0	–	–	–
Folic acid	0	–	0	+
C	–	+	+	0

Note: –, insensitive to hardly sensitive; 0, slightly sensitive to sensitive; +, very sensitive.

Table 5.5 Pelleting stability of vitamins [518].

Vitamin	Retention after 3 months of storage at RT (%)
A – conventional form	60 – 90
A – advanced form	80 – 90
D_3 – conventional form	70 – 90
D_3 – advanced form	80 – 100
E	95 – 100
K_3 (MSB)	50 – 70
B_1	60 – 80
B_2	95 – 100
B_6	80 – 100
B_{12}	50 – 80
Pantothenic acid	80 – 100
Niacin	95 – 100
Folic acid	50 – 80
Biotin	95 – 100
C	30 – 60

Table 5.6 Stability of vitamin C sources and other vitamins in an extruded catfish feed (retention, %) [518].

Vitamin	Feed stored (months)		
	0	1	3
Vitamin C source			
AASC	51	31	16
AAPP	100	100	96
Vitamin A		100	100
Vitamin E		94	100
Vitamin B1		100	100
Vitamin B2		100	100
Vitamin B6		100	100
Pantothenic acid		100	92
Folic acid		96	88
Biotin		84	87

Note: AASC = L-ascorbic acid; AAPP = L-ascorbyl polyphosphate.

leaching of water-soluble vitamins on prolonged immersion in water [519–520].

- Increased dietary vitamin requirement for optimum health and disease resistance, and optimum gut health, immunity, and stress control (over and above that required for optimum growth and feed efficiency) [1, 338, 358, 378, 394, 427, 434–435, 450, 483, 516].

- Possibility of complementary dietary interactions between individual vitamins and nutrients, and their consequent effect on specific dietary vitamin requirements, including the interactions between vitamin E and selenium [309–310], vitamin C and vitamin E [360, 472, 492, 508], vitamin C and folate [467], choline and methionine [525], choline and inositol [454], and choline and betaine [521].
- The possibility of satisfying specific dietary vitamin requirements through the endogenous synthesis of specific vitamins by the resident gut microflora of warmwater omnivorous and herbivorous cultured fish and crustacean species; depending upon the species, possibly vitamin K, vitamin B_{12}, biotin, folic acid, and/or inositol [110, 522–524].
- Finally, it also important to mention here that dietary vitamins may also exert toxic effects when present in excess, with high levels of Vitamin A causing skeletal abnormalities and scoliosis in Atlantic halibut (*Hippoglossus hippoglossus*) [149] and reduced growth and pale fragile livers in Japanese flounder (*Paralichthys olivaceus*) [335].

Anti-nutritional factors in feed ingredients

Anti-nutritional factors (ANFs) are secondary metabolites present in plant feed ingredients, which, if not eliminated and/or inactivated by processing, may exert a negative effect on fish and shrimp health and well-being [7, 110, 510, 512–514, 526–530, 683]. Although within plants these secondary metabolites serve protective functions against predation and plant survival [531], they can be broadly divided into four main categories, namely (1) proteins: protease inhibitors, phytohemagglutinins, toxic amino acids, food allergens; (2) glycosides: goitrogens, cyanogens, saponins, estrogens; (3) phenols: gossypol, tannins; and (4) miscellaneous: phytic acid, anti-vitamins, anti-enzymes, toxic fatty acids, and non-starch polysaccharides.

Table 6.1 shows the major ANFs reported within plant feed ingredients commonly used within aquaculture feeds. Notwithstanding the fact that soybean meal is the most commonly used dietary plant protein source within commercial aquaculture feeds [532], unprocessed soybean contains at least 14 different known endogenous ANFs, which in turn may may pose a risk to fish and shrimp health and well-being (Table 6.1).

6.1 Protease inhibitors

Protease inhibitors (PIs) are proteins that form complexes with individual protease enzymes (including trypsin, chymotrypsin, elastase,

Table 6.1 Reported endogenous anti-nutritional factors present in common plant feed ingredients.

Feed ingredient	Reported anti-nutritional factors	
Cereals		
Corn / maize *Zea mays*	Protease inhibitors	Phytoestrogens
	Phytic acid	Invertase inhibitor
	Cyanogens	
Rice *Oryza sativum*	Protease inhibitors	Lectins
	Phytic acid	Phytoestrogens
	Anti-thiamine factor	Lipase inhibitor
	Allergens	
Sorghum *Sorghum bicolor*	Protease inhibitors	Cyanogens
	Phytic acid	Tannins
	Amylase inhibitor	Lipase inhibitor
Wheat *Triticum vulgare*	Protease inhibitors	Lectins
	Phytic acid	Phytoestrogens
	Flatulence factor	Amylase inhibitor
	Dihydroxyphenylalanine	Lipase inhibitor
	Cyanogens	Allergens
Legumes		
Alfalfa/lucerne *Medicago sativa*	Protease inhibitors	Phytoestrogens
	Saponins	Anti-Vitamin E factor
Broad/faba bean *Vicia faba*	Protease inhibitors	Lectins
	Phytic acid	Tannins
	Dihyroxyphenylalanine	
Lupin seed meal *Lupinus albus*	Protease inhibitors	Saponins
	Phytoestrogens	Quinolizadine alkaloids
	Allergens	
Pea seed meal *Pisum sativum*	Protease inhibitors	Lectins
	Phytic acid	Cyanogens
	Saponins	Anti-Vitamin E factor
	Amylase inhibitors	
Oilseeds		
Cottonseed *Gossypium* spp	Phytic acid	Phytoestrogens
	Gossypol	Anti-Vitamin E factor
	Cyclopropenoic acid	Allergens
Linseed *Linum usitatissimum*	Cyanogens	Phytoestrogens
	Phytic acid	Anti-thiamine factor
	Anti-pyridoxine factor	
Mustard seed *Brassica juncea*	Protease inhibitors	Glucosinolates
	Anti-thiamine factor	Tannins
	Erucic acid	

Table 6.1 (Continued)

Feed ingredient	Reported anti-nutritional factors	
Peanut/groundnut *Arachis hypogaea*	Protease inhibitors	Lectins
	Phytic acid	Saponins
	Phytoestrogens	Lipase inhibitor
	Cyanogens	Allergens
Rapeseed *Brassica campestris napus*	Protease inhibitors	Glucosinolates
	Phytic acid	Tannins
	Erucic acid	Cyanogens
	Allergens	
Soybean *Glycine max*	Protease inhibitors	Lectins
	Glucosinolates	Saponins
	Phytic acid	Phytoestrogens
	Flatulence factor	Anti-Vitamin E factor
	Anti-Vitamin A factor	Anti-Vitamin D factor
	Anti-Vitamin B_{12} factor	Allergens
	Lipase inhibitor	Arginase inhibitor
Sunflower *Helianthus annus*	Protease inhibitors	Saponins
	Tannins	Arginase inhibitor
	Lipase inhibitor	Allergens

Note: Data compiled from numerous sources: [7, 110, 512–513, 561].

enterokinase and carboxypeptidase) and by so doing inhibiting their activity and digestive function within the gastrointestinal tract [110, 529]. It is generally believed that PIs are one of the most important ANFs present within soybeans (PIs of raw soybeans accounting for 6% of the total protein), and as such most commercial soybean processing methods have generally been targeted toward the destruction of PI activity, primarily through denaturation by thermal processing [513, 528, 533].

On a general basis dietary PIs have been found to reduce the apparent digestibility of both dietary protein and lipid, and increase the secretion of cysteine-rich pancreatic enzymes into the gastrointestinal tract; the latter possibly increasing the dietary requirement for sulphur-containing amino acids over and above that which would normally be required to remedy the normal methionine deficiency of soybean protein [110, 513, 529]. However, it is important to mention here that the destruction or inactivation of PIs, together with other heat-labile antinutrients (including other enzyme inhibitors, lectins, goitrogens, anti-vitamin factors, lathyrogens) is usually accompanied by a marked improvement in the nutritive value of plant protein sources for fish and shrimp [110, 513, 526, 529, 683].

6.2 Lectins

Lectins or phytohemagglutinins are soluble sugar-binding glycoproteins widely distributed in most plant feedstuffs; lectins in vitro causing the agglutination of vertebrate red blood cells [529]. Although reported toxicity signs of dietary plant lectins in humans and laboratory animals have ranged from reduced growth, alterations in gut microflora, to death (depending upon the toxicity of the specific lectin [512]), recent evidence suggests that lectins bind to intestinal epithelial cells, where they impair nutrient absorption and cause damage that may allow infiltration of bacteria into the blood stream [528, 683].

In salmonids, soybean lectins have been shown to bind to the distal intestine, and are believed to be involved in soybean meal-induced enteritis observed in these species (distal intestinal mucosa characterized by a shortening of the primary and secondary mucosal folds and a widening of the lamina propria, which is infiltrated by a mixed population of inflammatory cells identified as lymphocytes, macrophages, eosinophilic and neutrophilic granular cells, and diffuse immunoglobulin [529, 534]). Although lectins are usually reported as being heat-labile, their stability varies between plant species; many lectins being resistant to inactivation by dry heat and requiring the presence of moisture for more complete destruction [513, 535–536].

6.3 Saponins

Saponins are a heterogenous group of high molecular weight triterpenoid glycosides widely distributed in plants with the property of producing lather or foam when shaken with water; saponins being heat stable and usually removed from plants with hot water or alcohol [110, 512]. Toxicological studies with chicks have reported reduced growth and feed efficiency, and interference with the absorption of dietary lipids, cholesterol, bile acids, and vitamins A and E [537]. Moreover, saponins have been reported to affect the permeability of the mammalian intestinal epithelia, and by so doing facilitating the uptake of unwanted substances such as allergens and opportunistic pathogens and their toxins [110].

Dietary soybean saponins have also been reported to increase intestinal epithelial permeability in Atlantic salmon *Salmo salar* [538, 683], and are believed to play a key role in soybean-induced enteritis [529]. Similar negative alterations on the intestinal mucosal structure have also been reported in rainbow trout *Oncorhynchus mykiss* [539], gilthead seabream *Sparus auratam* [540] and the Japanese flounder *Paralichthys olivaceus* [541]. Finally, the ability of saponins to bind with cholesterol in the GI tract, raises concerns about the possible reduced availability of cholesterol for crustacean species and marine carnivorous fish species.

6.4 Phytoestrogens

Phytoestrogens are polyphenolic non-steroidal compounds with estrogen-like biological activity, and include isoflavonoids, flavonoids, stilbenes, and lignans; soybean containing high concentrations of isoflavones (as high as 4200 mg/kg) and being structurally similar to estradiol and mimicking its effects [110, 542–543, 683].

Due to their structural similarity to the hormone estradiol, one of the major concerns of phytoestrogens is their possible mimicking effects on reproduction [544–545]. Notwithstanding the above, studies using purified isoflavones have demonstrated a negative effect on the growth performance, intestinal function, liver metabolism and bone formation of Atlantic salmon *Salmo salar* fry [546].

6.5 Glucosinolates

Glucosinolates or goitrogens are thioglycosides commonly found in cruciferous plants (*Brassica* genus, including rapeseed and mustard seed), and also in soybean. Although glucosinolates are not harmful *per se*, on enzymatic hydrolysis with thio-glucosidases (myrosinase) they yield toxic products which are goitrogenic (that is, have antithyroid activity) and are also growth inhibitory, including goitrin, isothiocyanates, thiocyanate ions, nitriles, and vinyloxazolidinethiones. Goitrin is reported as being the most potent of the antithyroid agents and acts by inhibiting the thyroid gland to bind iodine; a process that cannot be reversed by dietary supplementation within iodine [512, 547–548, 683].

Reported deleterious effects of high glucosinolate ingestion in fish (either in purified form or through high dietary intakes of rapeseed meal and mustard seed) have been limited to reduced feed intake and growth, and thyroid hyperplasia [548–553]. Although thioglucosidase enzymes are readily destroyed by heat, glucosinolates are more heat resistant and may not always be completely destroyed. Moreover, although there is a possible risk of glucosinolate hydrolysis by gut microflora (and the consequent release of toxic products in the GI tract), glucosinolates are reported to be readily removed by extraction with hot water, dilute alkali, and/or organic solvent mixtures [110, 526].

6.6 Phytic acid

Phytic acid or the inositol hexaphosphate is the principal storage form of phosphorus and inositol within many plants, especially bran and seeds; forming chelates with mineral ions such as Ca, K, Mg, Zn, Cu, Fe and Mo resulting in these mineral elements being biologically unavailable to monogastric animals, including fish and shrimp. Table 6.2 shows average total phosphorus and phytate-phosphorus content of some common plant feed ingredient sources. Moreover, phytates are also reported to form sparingly soluble phytate-protein complexes in GI, and by so doing reducing dietary protein bioavailability [110, 512–513, 526, 554, 683].

The major reported deleterious effects of phytic acid have been mainly related to feed intake and reduced growth attributable mainly to the reduced availability of minerals, impaired protein digestibility, depressed thyroid function, and depressed absorption of nutrients due to damage of the intestinal epithelium [110, 526, 530]. Although the phytate content of plant meals can be significantly reduced by removing the outer endosperm or layer of seeds, the use of dietary exogenous enzyme supplements (phytase) and/or fermentation of ingredients prior to usage, have been found to be effective remedies for the removal and destruction of phytic acid [526, 530, 554–557].

6.7 Tannins

Tannins are a diverse group of water-soluble polyphenolic compounds which on hydrolysis yield either sugar residues and phenolic acids or non-hydrolyzable condensed tannins; common feed ingredients reported to contain considerable quantities of tannins include sorghum (containing up to 5% condensed tannins), copra (2.4% condensed tannins), sunflower seed meal (containing 1.2–2.7% chlorogenic acid), pea seed meal and the seed coat of brown legumes such as rapeseed and mustard seed [110, 512–513, 526, 683].

In terrestrial animals tannins have been reported to interfere with digestion by displaying anti-trypsin and anti-amylase activity and by binding to digestive enzymes or proteins and minerals, and have also been reported to reduce the absorption of vitamin B_{12} [512, 526]. Methods for the removal of tannins have included removal of the tannin-rich dark seed coat of legumes by dehulling, autoclaving, treatment with alkali, fermentation, and/or dietary supplementation with methyl donors such as choline and methionine to counteract the potential negative effects of chlorogenic acid in the case of sunflower seeds [110, 512, 526].

6.8 Cyanogens

Cyanogens are cyanogenic glucosides that release hydrogen cyanide on hydrolysis, and have been found in a variety of important food

Table 6.2 Total phosphorus and phytate-phosphorus content of common feed ingredient sources.

Ingredient	Total phosphorus (g/kg)	Phytate phosphorus (g/kg)	Phytate P: total P (%)
Barley	2.73	1.86	67.3
Canola meal	8.86	6.69	76.4
Coconut meal	4.30	2.40	56
Corn	2.50	1.70	73
Corn, gross defatted germ & bran	6.60	4.20	64
Corn, fine defatted germ & bran	12.10	7.80	65
Cotton seed whole	6.05	4.25	70.2
Cotton seed meal	11.36	9.11	80.5
Faba beans	4.01	2.43	60
Groundnut meal	6.00	4.60	77
Lupin *L. albus* whole	4.47	2.49	55.7
Lupin *L. angustifolius* whole	3.10	1.60	51.6
Lupin *L. angustifolius* dehulled	3.80	1.89	49.7
Maize	2.40	2.05	85.4
Maize gluten	5.00	4.20	84
Oats	2.43	2.10	86.4
Oats dehulled	2.25	1.15	51.1
Palm oil meal	5.10	2.90	57
Peas	3.45	1.67	48.4
Rapeseed meal	11.8	7	59
Rice	1.2	0.8	65
Rice broken	0.85	0.40	47.1
Rice bran	17.51	15.83	90.2
Rice polishings	15.7	11.3	72
Sorghum	2.92	2.41	82.7
Soybeans whole	5.55	3.08	55.5
Soybean meal	6.66	4.53	68.3
Sunflower meal	9.05	7.48	82.8
Wheat	3.08	2.20	74.9
Wheat by-products	8.02	7.00	87.3
Wheat bran	10.96	8.36	76.3

Note: adapted from [554].

plants, including barley, cassava, linseed, maize, peanut, peas, rapeseed, sorghum, and wheat. Moreover, although cyanogenic glucosides are generally heat stable and only sparingly soluble in water, within plants they are usually stored separately from their corresponding hydrolytic enzymes; the latter being readily destroyed by heat and cooking [110, 512].

Hydrogen cyanide is toxic to all life and animals through its ability to link with metalloproteins, and in particular cytochrome C oxidase of the mitochondrial respiratory chain, resulting in cessation of cellular respiration, cytotoxic hypoxia, lactic acidosis, and death [110]. Although Common carp *Cyprinus carpio* fed cyanogen-containing feed materials such as linseed meal and cassava have generally resulted in reduced growth, studies with Nile tilapia *Oreochromis niloticus* showed no difference in the growth performance of fish fed soaked sun-dried cassava leaf meal containing 9.9 ppm total cyanide with fish fed sun-dried cassava meal containing 71.1 ppm total cyanide [110, 526, 558–559].

6.9 Allergens

Plant allergens are generally heat-stable indigestible proteins that have the capability of stimulating antigenic effects or immunological activity, and capability of inducing gastrointestinal hypersensitivity, including intestinal mucosal lesions, specific and non-specific immune responses, and alterations in nutrient digestibility and reduced growth [110, 526, 560–561]. Common feed ingredients which have been shown to contain proteins which elicit antigenic activity in humans and animals (including fish), include soybeans (major antigenic proteins being glycinin and beta-conglycinin; these two proteins accounting for approximately 70–80% of total soy proteins; 564), cottonseed, groundnuts/peanuts, lupin, rapeseed, sesame, barley, rice and wheat [561, 683].

Although some authors have demonstrated that high levels of immunologically active glycinin and beta-conglycinin in different soy preparations seemed to cause inflammation and negatively affect growth performance in rainbow trout (*Oncorhynchus mykiss*; [562]), other authors working with the same species reported no antigenic activity in soy protein concentrate, although glycinin and beta-conglycinin was detected in defatted toasted soy flour [563]. Solid-state fermentation of soybean meal with *Bacillus subtilis* has been found to be an effective method for reducing the trypsin inhibitor activity and beta-conglycinin content of soybean meal [564].

6.10 Oligosaccharides and non-starch polysaccharides

Oligosaccharides and non-starch polysaccharides (NSP) are important constituents of legumes seeds and grains; defatted soybean meal containing 12–15% soluble carbohydrates (6–7% sucrose, 1–2% raffinose and 5–6% stachylose) and 14–18% NSPs (arabinan, arabinogalactan and acid polysaccharides), whereas in grains NSPs consist primarily of cellulose, hemicellulose and lignin [526]. Although not toxic *per se*, oligosaccharides (includes raffinose and stachylose) and NSPs are largely indigestible by most monogastric animals, their negative effects in fish have been ascribed to their lower energy value (compared to more digestible carbohydrates such as glucose and starch), their ability to bind with bile acids and hinder the action of digestive enzymes and nutrient assimilation, alter the composition of the gut microbiota (including the *in-situ* fermentation of flatulence causing oligosaccharides), as well as affecting gut passage time [526, 529–530].

In general, soluble NSPs are believed to be more detrimental to fish growth than oligosaccharides and insoluble NSPs (structural polysaccharides: cellulose, hemicellulose and lignin), due to their ability to trap water and form gum-like masses in the intestine, and by so doing possibly negatively affecting gut viscosity and hindering digestive enzyme action and nutrient assimilation [565]. Structural NSP levels (often referred to as crude fiber) can be reduced either by physical processing (dehulling seeds or by air classification), oligosaccharides removed by alcohol/aqueous extraction, and NSPs and oligosaccharides reduced and/or eliminated through microbial fermentation and/or through the use of specific dietary exogenous enzymes [530, 556, 566–567].

6.11 Miscellaneous anti-nutritional factors

Gossypol

Gossypol is a lipid-soluble polyphenolic compound present in the pigment glands of cottonseed, with levels as high as 6% reported within glanded cottonseed to as low as 0.01% within new glandless varieties. Reported toxic effect of gossypol in fish vary within and between species, and have included reduced feed intake, reduced growth, hematocrit, hemoglobin, and reproductive capacity, with histological lesions reported in the liver, kidney, spleen and gonads [110, 568, 683].

Erucic acid

Erucic or *cis*-13-docosenoic acid (20:1n-9) is a fatty acid present in plants belonging to the Brassica family, with levels as high as 60% reported in some rapeseed and mustard seed oils. Although erucic acid has been reported to be cardiotoxic in some animals, negative effects of high erucic acid rapeseed oil have been observed in coho salmon fed diets with 6% and 12% rapeseed oil, including reduced growth, increased mortality, and histopathology in the skin, gills, kidney, and heart [110, 683].

Alkaloids

Alkaloids represent a large and structurally diverse group of secondary plant metabolites, and include nicotine (tobacco), atropine (deadly nightshade), cocaine (coca plant), quinine (cinchona bark), and morphine (opium poppy). However, the only seed legume likely to be used within aquafeeds which contains alkaloids, in the form of quinolizidine alkaloids, is the lupin *Lupinus* ssp. Reported toxicity signs in fish have included reduced feed palatability and growth [110, 526].

Cyclopropene fatty acids

Cyclopropene fatty acids such as sterculic acid and malvalic acid are present in cottonseed oil and meal, and in fish are believed to have a negative effect on growth and lipid metabolism, including histological abnormalities such as hepatocycte necrosis, unusual liver glycogen deposition, appearance of 'fibres' in the hepatocyte cytoplasm, proliferation of bile ducts, and fibrosis [569, 683].

Toxic amino acids

Numerous natural occurring toxic amino acids have been identified within plant feedstuffs, including (but not limited to): (1) *Mimosine* found in the leaves of ipil-ipil tree *Leucaena leucocephala*; (2) *Canavanine* found in the legume sesbania *Sesbania* spp. and Jack bean *Canavalia* spp.; (3) *Dihydroxyphenylalanine* found in the faba or broad bean *Vicia faba*; (4) *Methylselenocysteine* and *selenocystathionine* found in selenium accumulating plants *Astragalus* and *Machaeranthera* spp.; and (5) *Indospicine* found in the creeping indigo *Indigofera spicata* [110, 512–513, 683]. The majority of these toxic amino acids have been reported to have a negative effect on fish growth and feed efficiency.

Anti-vitamins

Numerous secondary metabolites in plants have been reported which exhibit anti-vitamin activity, including: (1) anti-thiamine factor (thiaminase) present in cottonseed, linseed, mung bean; (2) anti-vitamin A, D, E and B_{12} factors present in soybeans; (3) anti-vitamin E factor present in kidney beans, soybeans, alfalfa, and field pea; (4) anti-vitamin K factor present in sweet clover; (5) anti-niacin factor present in sorghum; and (6) anti-pyridoxine factor present in linseed [512–514, 683].

Anti-enzymes

Anti-enzyme factors have been reported in many major plant feed ingredients, including: (1) anti-PIs in corn, rice, sorghum, wheat, lupin, pea seed meal, mustard seed, peanut, rapeseed, soybeans, and sunflower seed; (2) anti-amylase inhibitor in sorghum, wheat, and peas; (3) anti-invertase inhibitor in corn; (4) anti-arginase inhibitor in sunflower seeds and soybeans; (5) anti-cholinesterase inhibitor in potato proteins: and (6) anti-lipase activity in rice, sorghum, peanut, soybean, and sunflower seed (Table 6.1) [512–513, 683].

Table 6.3 Antinutrients and treatment methods employed for their removal or reduction.

Anti-nutritional factor	Feed source	Treatment methods used
Amylase inhibitors	Cereal grains, peas	Heat
Fiber	All plants	Dehulling
Goitrogens	Rapeseed	Heat, Iodine supplementation
Gossypol	Cottonseed	Non-polar extraction, iron supplementation
Protease inhibitors	Cereal grains, legumes	Heat, methionine supplementation
Lectins	All plant seeds	Heat, supplementation with specific carbohydrates
Lipase inhibitors	Cereal grains, beans	Heat
Oligosaccharides	Legumes	Alcohol/aqueous extraction
Phytic acid	All plants	Mineral supplementation
Phytoestrogens	Beans	Alcohol/non-polar extraction
Phytosterols	Legumes	Alcohol/non-polar extraction, cholesterol supplementation
Quinolozidine alkaloids	Lupins	Aqueous extraction
Saponins	Legumes	Alcohol extraction
Tannins	Rapeseed, beans, sorghum	Dehulling, autoclaving

Notes: For all of the antinutrients listed, levels may be reduced by fermentation or through the use of dietary exogenous enzymes that specifically inactivate the compound, or through the use of selective breeding techniques or genetic modification [modified after 514, 526, 528, 556].

6.12 Processing methods for removal

In conclusion, Table 6.3 summarizes the reported major processing methods used to eliminate and/or reduce the level of specific ANFs within different plant feed ingredient sources [modified after 514, 526, 528, 556].

On a final note, it should be pointed out that more knowledge and understanding is needed concerning the possible interactions between individual ANFs on fish and shrimp health and well-being, and in particular for those feed ingredients known to contain multiple ANFs; soybean is reported to contain up to 14 different endogenous ANFs (Table 6.1).

7

Feed contaminants

Last but not least, feed ingredients may be contaminated with adventitious toxic substances (depending upon their origin, handling and/or processing before use) which may have a negative effect on fish and shrimp growth, and/or health and well-being, including the following.

- **Mycotoxins** are present in most processed plant feed ingredient sources, and generally believed to be the most important feed contaminant and present the highest risk factor for all cultured species (mycotoxins will be discussed in greater detail later in this section [110, 530, 570–571, 574–575, 683]).
- **Environmental chemical contaminants** may be present in some aquatic and terrestrial animal proteins and lipids, and includes metals/metalloids (mercury, lead, cadmium, arsenic), polycyclic aromatic hydrocarbons, persistent organic pollutants, perfluorinated compounds (PFCs), pharmaceutical and personal care products, radioactive elements (caesium-137, iodine-131, strontium-90), and micro-plastics. They present a low to medium risk factor for carnivorous fish species [110, 530, 656–659, 682–683].
- **Pesticides and herbicide residues** may be present in plant feed ingredients, and includes glyphosates and their breakdown products – low to medium risk factor for all species [660–663];

- **Veterinary drug residues** may be present in terrestrial animal protein and lipid sources, and includes antibiotics, hormones, and growth promoters. They present a low to medium risk factor for all cultured species [664–668, 682–683].
- **Miscellaneous** substances, excludes oxidized lipids, and may include adulterants (melamine) [669–672], transmissible spongiform encephalopathies (prions – ruminant protein sources) [673–676], biotoxins [683], and the possible use of transgenic feed ingredient sources (due to potential human food safety concerns [530, 677–681, 683]).

7.1 Mycotoxins

Risks

Mycotoxins are toxic secondary metabolites produced by saprophytic filamentous fungi or molds, and in particular to molds belonging to the genera *Aspergillus*, *Penicillium* and *Fusarium* (Figure 7.1). To date more than 400 different mycotoxins have been identified, from a total of over 1.5 million mold species [570–575].

For the purpose of this review, the major mycotoxins of concern to human health, the food sector, and the animal and aquaculture feed manufacturing sector are:

- aflatoxin (AFL): major metabolites – AFB_1, AFB_2, AFG_1, AFG_2
- fumonisin (FUM): major metabolites – fumonisin B1, B2, B3
- ochratoxin A (OTA): major metabolite – ochratoxin A
- trichothecenes (TCT); major metabolites – T-2 toxin, deoxynivalenol (DON), nivalenol
- zearalenone (ZEN); major metabolite – zearalenone.

Apart from the deleterious effects of mycotoxins on human health following direct inhalation and/or ingestion [571, 573,

Acetoxyscirpenediol	Acetyldeoxynivalenol	Acetylneosolaniol	Acetyl T-2 toxin	Aflatoxin	Aflatrem	Altenuic acid
Alternariol	Austdiol	Austamide	Austocystin	Avenacein +1	Beauvericin +2	Bentenolide
Brevianamide	Butenolide	Calonectrin	Chaetoglobosin	Citrinin	Citreoviridin	Cochliodinol
Crotocin	Cytochalasin E	Cyclopiazonic acid	Deacetylcalonectrin	Deoxynivalenol diacetate	Deoxynivalenol monoacetate	Diacetoxyscirpenol
Destruxin B	Enniatins	Fructigenin +1	Fumagilin	Fumonisin B1	Fusaric acid	Fusarin
Gliotoxin	HT-2 toxin	Ipomeanine	Islanditoxin	Lateritin +1	Lycomarasmin +1	Malformin
Maltoryzine	Moniliformin	Monoacetoxy scirpenol	Neosolaniol	Nivalenol	NT-1 toxin	NT-2 toxin
Ochratoxin	Oxalic acid	Patulin	Penicillic acid	Penitrem	Roridin E	Rubratoxin
Rubroskyrin	Rubrosulphin	Rugulosin	Sambucynin +1	Satratoxins FGH	Scirpentrio	Slaframine
Sterigmatocystin	T-1 toxin	T-2 toxin	Triacetoxy scirpendiol	Trichodermin	Trichothecin	Trichoverrins
Trichoverrols	Tryptoquivalene	Verrucarin	Verruculogen	Viopurpurin	Viomellein	Viriditoxin
Xanthocillin	Yavanicin+1	Zearalenone	Source: https://www.mold-help.org/mycotoxin list			

Fusarium SP	Alternaria SP	Monpgraphella Sp	Chaetoglobosin SP	Stachybotrys SP
Aspergillus SP	Myrothecium SP	Acremoniun sp	Rhizoctonia sp	Trichoderma sp
Penicillium sp	Trichothecium sp	Trichophyton sp	Eurotium sp	

Figure 7.1 Wide diversity of mycotoxins produced by molds: 300 to 400 metabolites to date.

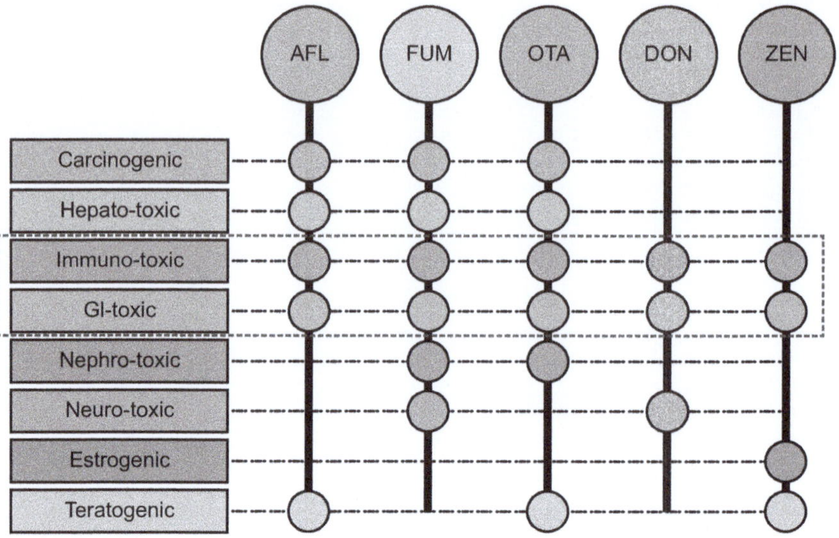

Figure 7.2 Major reported health effects of aflatoxin (AFL), fumonisin (FUM), ochratoxin A (OTA), deoxynivalenol (DON), and zarelonone (ZEN) in humans and animals (modified after [571]).

575–579] (Figure 7.2), it has been estimated that over 25% of food crops around the world are contaminated with mycotoxins [580]. Moreover, according to the latest 2019 BIOMIN world mycotoxin survey (93,543 analyses conducted on 21,287 finished animal feeds and raw commodity samples from 86 countries), the most prevalent mycotoxins were FUM (70%) and DON (68%) [581].

In the case of aquaculture, Table 7.1 shows the reported mycotoxin contamination of some plant ingredients and finished aquaculture feeds in Asia (Asia accounting for over 91.7% of total global aquaculture production in 2018 [582]). Of particular note is the fact that most feed ingredients and finished feeds are usually contaminated with multiple mycotoxins, and that these mycotoxin levels tend to be concentrated and higher within plant by-product meals than in the original seed (depending upon on the seed processing method employed – such as the production of DDGS and corn gluten meal from whole corn or the production of wheat bran from whole wheat [579, 583]).

Table 7.1 Mycotoxin contamination of plant feed ingredients and finished aquaculture feeds in Asia.

Ingredient	AFL N/P% X/max	ZEN N/P% X/max	DON N/P% X/max	FUM N/P% X/max	OTA N/P% X/max
Plant meals	2316/29%	2327/64%	2324/81%	2301/80%	2270/15%
sampled[a]	**49/4,890**	**150/9,573**	**671/53,796**	**1,507/123,444**	**7/124**
Soybean	71/18%	62/19%	65/14%	69/6%	61/26%
meal[b]	**0/3.4**	**4/65.3**	**25/840**	**22/866**	**1/11.7**
Rapeseed	10/20%	15/20	15/47%	10/0%	10/30%
meal[b]	**0/0**	**2/13.5**	**121/685.2**	**0/0**	**0/3.1**
Corn[a]	353/31%	353/61%	353/84%	346/99%	328/4%
	51/605	**243/5,087**	**611/7,632**	**2,786/47,485**	**6/26**
DDGS[a]	39/18%	39/95%	39/95%	39/95%	39/15%
	21/39	**242/1,712**	**1,994/7,368**	**5,378/43,610**	**5/15**
Corn gluten	17/59%	17/94%	17/94%	15/100%	15/20%
meal[a]	**130/441**	**2,407/9,573**	**989/8,194**	**13,686/60,275**	**15/17**
Wheat	172/5%	341/31%	434/70%	189/10%	186/13%
flour[b]	**0/11.1**	**38/2,991**	**842/12,000**	**33/2,273**	**0/30**
Rice bran[a]	27/63%	27/78%	27/59%	27/63%	27/19%
	8/34	**95/691**	**112/580**	**275/1,742**	**4/13**
Finished	1081/39%	1089/80%	1089/92%	1081/96%	1072/20%
feeds[a]	**26/697**	**78/2,089**	**331/6,126**	**1,103/27,352**	**7/124**

Note: Where N/P% indicate number of samples tested, % contaminated samples, and X/max indicate the mean and maximum concentration of the mycotoxin within the positive samples, expressed as ug/kg or ppb; [a][584]; [b][585].

In general, crop mycotoxin contamination will *depend* upon numerous interrelated factors and interventions during the crop production and handling process [110, 571, 586–587], namely:

- during the crop production cycle, *depending* upon plant species and variety, soil conditions and fertilizer use, irrigation and weather, length of the production period, susceptibility of the crop to diseases and damage from insects and birds, and use of herbicides, fungicides, and pesticides
- during the harvesting of the crop, *depending* upon stage of crop maturity, weather at the time of harvesting, harvesting method, speed of harvesting, and necessity of mechanical and/or chemical processing/treatment prior to storage

- during the storage of the processed crop by the farmer or distributor, *depending* upon method of storage, packing material (if any), length of storage, and environmental storage conditions – temperature, relative humidity
- during the transportation of the processed crop to the feed manufacturer and/or end user, *depending* upon the transportation method employed, method of packaging and handling, protection from the elements and pests, and length of transportation
- during the storage and use of the processed crop by the feed manufacturer, *depending* upon the feed ingredient storage conditions, need for further processing (grinding) before use, feed formulation employed (use of mycotoxin binders, organic acids), feed processing method employed (grinding, pelleting, extrusion), and storage conditions/packing of the finished feed prior to distribution to the farmer
- during the on-farm storage, handling, and use of the finished feed by the farmer, *depending* upon the on-farm feed storage conditions employed, length of storage before use, environmental storage conditions employed – temperature, relative humidity, need for further feed processing/top-dressing before use (such as the addition of probiotics and feed additives), and feeding method employed.

Impacts

The impact of mycotoxins are evident in four main areas: (1) their impact on crop losses due to mold disease; (2) their impact on the reduced commercial value of mycotoxin contaminated food and feed; (3) their impact on human and animal health due to inhalation and/or consumption; and (4) the cost of mycotoxin mitigation and control strategies [571]. For the purpose of this review, only the reported health impacts of the major key mycotoxins on fish and shrimp health will be reported here (Table 7.2).

Although the reported health impacts will vary depending upon the level and source of the mycotoxin tested, the duration of the feeding trial, and the developmental status of the test animal (larvae, juvenile, grow-out, brood), some generalizations can be made, namely (1) multiple mycotoxins are usually present within single feed ingredients and finished feeds, with corn products usually being the most contaminated [579], (2) mycotoxins usually exert a negative effect on fish or shrimp growth, feed conversion efficiency, and survival [588], (3) mycotoxins usually exert a negative effect on the immune system and disease resistance [576, 605, 620–622], and mycotoxins usually exert a negative effect on gut health, and intestinal integrity and permeability [577, 605, 624].

Possible solutions

The risks of mycotoxin contamination and mycotoxicosis are real and of paramount economic importance. Sadly, most feed manufacturers and farmers are not aware of this, and the consequent profound effect of mycotoxins on the health and well-being of farmed fish or shrimp [576]. To a large extent this has been due to the difficulty of rapidly measuring specific mycotoxin levels within feed ingredients and finished feeds (including associated costs, training and equipment requirements [625–626]), and secondly because of the difficulty of proving that sub-optimal growth, feed efficiency or survival is the direct result of mycotoxicosis and not the result of other dietary factors, including the presence of other possible dietary ANFs [110, 529, 586].

The possible solutions or remedies to mycotoxin mitigation are as follows.

- Limit the use of lower cost and lower quality feed ingredient sources likely to be contaminated with mycotoxins, such as cereal by-product meals, the use of expired feed ingredients, and the use of floor sweepings.
- Use of dietary organic acids (and their derivatives) and other feed additives to prevent and/or limit mold growth within finished feeds on prolonged storage, including propionic acid, sorbic acid, herbs and spices, essential oils, phenolic antioxidants, etc. [627–632].
- Use of dietary mycotoxin binders to prevent the absorption of mycotoxins by the target species, including activated

Table 7.2 Reported major health impacts of key mycotoxins in farmed fish and shrimp.

Aflatoxin (AFL B_1 B_2 G_1 G_2) Major reported health impacts: – Reduced growth, feed efficiency, and survival; – Increased incidence of liver cancer, necrosis, liver lesions; – Yellowing eyes, mucous membranes and skin (jaundice); – Immuno-suppression and reduced disease resistance; – Reduced erythrocyte, leucocyte, hemoglobin count; – Reduced hematocrit, impaired blood clotting, pale gills; Useful references: Biomin: Mycotoxins in aquafeed: a hidden threat to fish and shrimp farming (https://www. biomin.net/species/aquaculture/mycotoxins-in-aquaculture/) [110, 579, 586, 588–596].	Major species studied: Atlantic salmon, Black tiger shrimp, Channel catfish, European seabass, Gibel carp, Gilthead seabream, Nile tilapia, Pacific white shrimp, Pangasius catfish, rainbow trout, Red drum, Rohu, Sturgeon, Yellow catfish; Common AFL contaminated ingredients: cassava, copra (coconut), corn (maize), corn gluten meal, cottonseed meal, dried distillers' grains with solubles, groundnut (peanut) meal, palm kernel meal, soybean meal, sunflower seed meal, rice bran, wheat, wheat bran;
Fumonisin (FB_1, FB_2, FB_3) Major reported health impacts: – Reduced growth, feed efficiency, and survival; – Liver, kidney and pancreas lesions; – Low hematocrit values; – Endothelial permeability; – Brain lesions and necrosis; – Immune-suppression; – Reduced disease resistance; Useful references: Biomin: Mycotoxins in aquafeed: a hidden threat to fish and shrimp farming (https://www. biomin.net/species/aquaculture/mycotoxins-in-aquaculture [110, 579, 584, 586, 588–590, 593, 597–600, 619]).	Major species studied: African catfish, Atlantic salmon, Channel catfish, European seabass, Common carp, Gilthead seabream, Nile tilapia, Pacific white shrimp, Pangasius catfish, rainbow trout; Common AFL contaminated ingredients: corn (maize), maize flour, corn gluten meal, dried distillers' grains with solubles, soybean meal, rice bran, wheat, wheat flour, wheat bran;

Ochratoxin A (OTA)

Major reported health impacts:

– Reduced growth, feed efficiency, and survival;
– Liver and kidney lesions & necrosis;
– Pale, swollen kidneys;
– Physical deformities of head, tail & eyes;
– Reduced disease resistance;

Useful references: Biomin: Mycotoxins in aquafeed: a hidden threat to fish and shrimp farming (https://www. biomin.net/species/ aquaculture/mycotoxins-in-aquaculture [110, 579, 586, 588–590, 593, 601–604])

Major species studied: Atlantic salmon, Channel catfish, European seabass, Common carp, Gilthead seabream, Nile tilapia, Rainbow trout;

Common AFL contaminated ingredients: corn (maize), corn gluten meal, cottonseed meal, dried distillers' grains with solubles, groundnut (peanut) meal, palm kernel meal, soybean meal, sunflower seed meal, rice bran, wheat, wheat bran;

Trichothecene (TCT) – Deoxynivalenol (DON)

Major reported health impacts:

– Reduced growth, feed efficiency, and survival;
– Reduced hematocrit;
– Liver histopathology changes;
– Reduced disease resistance;
– Immune-suppression;

Useful references: Biomin: Mycotoxins in aquafeed: a hidden threat to fish and shrimp farming (https://www. biomin.net/species/aquaculture/mycotoxins-in-aquaculture [110, 579, 586, 588–589, 593, 605–611])

Major species studied: Atlantic salmon, Channel catfish, Common carp, Pacific white shrimp, rainbow trout;

Common AFL contaminated ingredients: barley, corn (maize), maize flour, corn gluten meal, groundnut (peanut) meal, dried distillers' grains with solubles, soybean meal, rice bran, wheat, wheat flour, wheat bran;

Table 7.2 (Continued)

Trichothecene (TCT) – T-2 toxin Major reported health impacts: – Reduced growth, feed efficiency, and survival; – Reduced hematocrit & hemoglobin levels; – Tail formation abnormality; – Immuno-suppression; – Reduced disease resistance; – Hepatopancreas & kidney abnormalities; – Increased sensitivity to oxidative stress; Useful references: Biomin: Mycotoxins in aquafeed: a hidden threat to fish and shrimp farming (https://www. biomin.net/species/aquaculture/mycotoxins-in-aquaculture [110, 579, 586, 588–590, 593, 604, 612–614])	Major species studied: Channel catfish, Common carp, Pacific white shrimp, rainbow trout; Common AFL contaminated ingredients: corn (maize), corn gluten meal, dried distillers' grains with solubles, wheat, wheat flour, wheat bran;
Zearalenone (ZEN) Major reported health impacts: – Reduced growth; – Reproductive disfunction & disorders; – Macrocytic hypochromic anemia; – Reduced hematocrit & hemoglobin levels; – Immuno-suppression; – Reduced disease resistance; Useful references: Biomin: Mycotoxins in aquafeed: a hidden threat to fish and shrimp farming (https://www. biomin.net/species/ aquaculture/mycotoxins-in-aquaculture [110, 579, 586, 588–590, 593, 615–618])	Major species studied: Black tiger shrimp, European seabass, Common carp, Rainbow trout; Common AFL contaminated ingredients: cassava, corn (maize), corn gluten meal, cottonseed meal, dried distillers' grains with solubles, fishmeal, groundnut (peanut) meal, palm kernel meal, rice bran, wheat, wheat flour, wheat bran;

charcoal, inorganic binders (clay minerals, aluminosilicates, bentonite, montmorillonite, smectite, hydrated sodium calcium aluninosilicate, zeolite), organoclays (organophilic clays, organozeolite, montmorillonite with organic compounds, bentonite with quaternary amines), organic binders (yeast, yeast cell wall extracts, mannan oligosaccharides, glucomannans, peptidoglycans, beta-glucans, algae extracts, chitin oligosaccharides), and synthetic polymers (cholestryamine, polyvinylpyrrolidone [577, 602, 621, 633–644]).

- Use of microorganisms or enzymes that can biotransform mycotoxins in the gastrointestinal tract [577, 645–648]. It is important to mention here that the fermentation process can also be used to biotransform and reduce mycotoxin levels [571, 649–652].
- At the farm level, to ensure that finished feeds are stored under cool well-ventilated conditions, are used on a first-in, first-out basis, are not subjected to adulteration by top-dressing prior to usage, and that feed bags are not left outdoors for prolonged periods of time (hours) prior to feeding, following recommended best aquaculture feed management practices [587, 653–654].

On a final note, it must be emphasized that mycotoxins are extremely toxic substances that pose a serious health risk to humans, and in the case of aquaculture this concerns the potential risk of mycotoxin carry-over into aquaculture food products [590–591, 572]. Bearing this is mind, Table 7.3 summarizes the current mycotoxin regulations for animal feed (including fish feed) and selected feed ingredients within the European Union and the USA [623].

Table 7.3 Mycotoxin regulations for animal feed and selected ingredients in the EU and USA.

EU		USA	
Aflatoxin in feed – All feed materials – Complementary & complete feed	Max level (B_1) 20 ppb 10 ppb	**Aflatoxin in feed** – Corn, corn products, peanut products, cottonseed meal, and other animal feeds & feed ingredients	Max level (total) 20 ppb
Deoxynivalenol in feed – Cereals and cereal products – Maize by-products – Complementary and complete feeding stuffs	Guidance level 8 ppm 12 ppm 5 ppm	**Deoxynivalenol in feed** – Grain and grain by-products destined for other animals	Guidance level 5 ppm
Fumonisin in feed – Complementary & complete feeding stuffs for fish	Guidance level (B1 & B2) 10 ppm	**Fumonisin in feed** – Corn & corn by-products for catfish – All other species	Guidance level (B1, B2, B3) 20 ppm 10 ppm
Ochratoxin A – Cereals & cereal products – Complementary & complete feeding stuffs for pigs & poultry	Guidance level 0.25 ppm 0.05 ppm 0.10 ppm	No guidance given	
T-2 & HT-2 Cereal products for feed & compound feed: – Oat milling products (husks) – Other cereal products – Compound feed	Guidance level 2000 ppb 500 ppb 250 ppb	No guidance given	

Source: https://www.romerlabs.com/en/knowledge-center/knowledge-library/articles/news/worldwide-mycotoxin-regulations/.

8

Concluding remarks

The nutritional pathology conditions described in this book can be summarized under the following major headings.

- External pathology signs and abnormalities
  - **Animal behavior:** abnormal/erratic swimming (ataxia); lethargy; irritability; spasms; convulsions; shock syndrome; increased cannibalism and aggressive behavior; disrupted schooling behavior; hydrops (floating bodies); sluggishness; loss of equilibrium; photophobia; reduced stress response; reduced oxidative stress response; reduced handling stress.
  - **Body form, shape, and size:** increased body deformity (short body); short-body dwarfism; lordosis; scoliosis; opercula deformities; jaw deformities; snout deformities; ascites; atrophy; abnormal tail growth; caudal fin deformities; truncated snout; abdominal swelling; yellow ascitic fluid in abdomen; marked congestion, with severe hemorrhage, in dermal vessels around snout and base of fins; steatitis of all abdominal fat; subcutaneous edema; pericardial edema; branchial congestion; foreshortened maxillae; hydrocephalus; droopy tail syndrome.
  - **Eyes, fins, and gills:** clubbed gills; cataract; increased eye abnormality; reduced visual organ development; eye lesions; lens cataracts; exophthalmia; increased fin erosion and loss; pale gill color; impaired gill structure integrity and

antioxidant capacity; caudal fin erosion; erosion of caudal peduncle; pale gills; flipped gill cover; needles and plate-like melamine-cyanurate crystals with multifocal hemocytic granulomas in the antennal gland tubules; orbital collapse; reduced gill immunity; impaired gill antioxidant system and tight junction proteins; severe hyperplasia of epithelial cells of gill lamellae; red fins; rapid opercular movements.

- **Feed intake and feed efficiency:** reduced feed intake (anorexia); reduced growth; reduced feed efficiency.
- **Lesions and hemorrhages:** skin hemorrhage; fin hemorrhage; mouth lesions; gill lesions; ulcers; parasite infestation; hemorrhagic exophthalmia; reduced wound repair; hemorrhaging operculum, lower jaw, isthmus and pelvic fin.
- **Skin/exoskeleton color and pigmentation:** depigmentation; lighter tail skin color; yellow skin and gill pigmentation; pale body color; loss of body coloration; dark or light skin pigmentation; lack of skin color; dark coloration; increased scale loss; reduced molting capability; reduced shell hardening; soft shell; incomplete molting; desquamation of the skin; thinning of epidermis with a decrease of mucous cells.

Internal pathology signs and abnormalities

- **Blood and hematology:** anemia; reduced erythrocyte/red blood cell count; reduced hematocrit value; reduced hemoglobin value; reduced packed cell volume; reduced leucocyte count; increased erythrocyte osmotic fragility; reduced erythrocyte volume; hemolytic anemia; microcytic anemia; increased hemolysis; increased blood clotting time; hyperchromic anemia; exudative diathesis; aniosocytosis; poikilocytosis; leukocytosis and decrease in neutrophils and lymphocytes.
- **Gut structure and function:** reduced nutrient digestibility; reduced digestive capacity and functionality; reduced digestive function and gut microbiota; reduced intestinal digestive enzyme activity; increased intestinal gas

production; reduced lipase and antioxidant enzyme level; impaired intestinal structure and functionality; reduced intestinal morphology and function; impaired intestinal immune response; reduced intestinal health and antioxidant capacity; reduced digestive and antioxidant capacity; reduced intestinal immune response, tight junctions and microflora; reduced digestive capacity, microflora and immune response; destruction of intestinal mucosal epithelium; increased intestinal oxidative stress; reduced intestinal health and concentration of goblet cells in foregut; reduced intestinal fold height; reduced intestinal mucosal immune and intestine physical function; damaged gut mucosa; periarteritis in submucosa layer of intestine; yellow-brown intestinal contents; inflammation and dysfunction of intestinal barrier.

- **Heart:** increased heart weight and atherosclerotic alterations; pericarditis and endocarditis; heart myopathy; enlarged heart; enlarged heart with hemorrhage in the epicardium; heart edema.
- **Immunity and disease resistance:** reduced disease resistance; reduced immune response; reduced innate immune function; reduced non-specific immune function.
- **Kidney/spleen structure and function:** melamine-cyanurate crystal deposits in renal tubules; kidney granulomatosis; enlarged spleen and kidney; hyperplasia of renal hematopoietic tissue; ceroid deposition in the spleen; sclerotic glomeruli in posterior kidney and ceroid-laden macrophages in intrahepatic blood vessels; swollen kidney with hyperemia and hemorrhage; splenic hemosiderosis; shrunken glomeruli and hepatocytes and occurrence of yellow ceroid pigments; pale yellow spleen.
- **Liver/hepatopancreas/gall bladder structure and function:** swollen pale liver; fatty liver; liver degeneration; green liver; liver granulomatosis; increased hepatosomatic index; liver lipoid degeneration and ceroid accumulation within hepatocytes; yellow-orange liver coloration; liver ceroid accumulation; hepatic lesions; swollen yellow/

blue colored liver; increased liver lipid accumulation; lipid malabsorption syndrome; extensive liver diffused fat vacuolation and fatty cyst formation; enlarged gall bladder; reduced liver antioxidant activity; bile acid accumulation; jaundice disease; marked distention of bile duct; pale foamy appearance of the hypervacuolated enterocytes of the pyloric caeca;

- **Muscle structure and function:** skeletal muscle degeneration; muscular dystrophy; degeneration of skeletal and cardiac muscle fibers; extreme atrophy and necrosis of white muscle fibers; Thin back disease; Sekoke disease; marked loss of flesh on the back; apparent muscular weakness; lateral muscle myopathy; severe muscle damage.
- **Reproductive function:** reduced sperm quality and motility; reduced reproductive performance and larval quality; reduced maturation; reduced reproductive performance and egg quality.
- **Skeletal and bone abnormalities:** bone deformities; spinal deformities; lordosis; scoliosis; vertebral curvature; twisted gill filaments; dislocated vertebrae; vertebral lesions; compressed spinal cord; isthmus breakage; degenerative changes in thoracic vertebrae; deformed vertebrae; reduced skeletal ossification; reduced bone mineralization; delayed ossification of bone endochondral structures; broken-skull disease; increased larval deformity; gill filament alterations; distortion of gill filament cartilage; hyperplasia, hypertrophy and dysplasia of the bone cartilage of gill filaments; edematous separation and telangiectasia of gill lamellae.
- **Survival and mortality:** reduced survival; increased mortality; and death.

Table 8.1 summarizes the major nutritional pathology conditions reported in farmed fish and shrimp, and the corresponding dietary nutritional imbalances that caused them.

Last, but not least, it is important to mention that one of the most profound pathological effects of dietary nutrient deficiencies,

Table 8.1 Major nutritional pathology conditions and reported causes in farmed fish and shrimp.

Major pathology sign	Nutrient/ANF/ contaminant	Species
Animal behavior: ataxia	Thiamine – deficiency	Japanese eel, rainbow trout
	Pyridoxine – deficiency	Spotted snakehead
Hyperirritability	Riboflavin – deficiency	Barramundi
	Pyridoxine – deficiency	Gilthead seabream, orange-spotted grouper, spotted snakehead
Muscle spasms	Pyridoxine – deficiency	Hybrid red tilapia, spotted snakehead
Convulsions	Potassium – deficiency	Chinook salmon
	Magnesium – deficiency	Common carp
	Vitamin E – deficiency	Rainbow trout
	Thiamine – deficiency	Barramundi
	Pyridoxine – deficiency	Hybrid red tilapia, Indian catfish, orange-spotted grouper, spotted snakehead
Erratic swimming behavior	Biotin – deficiency	Indian catfish, Nile tilapia, walking catfish
	Pyridoxine – deficiency	Gilthead seabream
	Vitamin C – deficiency	Giant tiger prawn, spotted rose snapper
Sluggishness	Magnesium – deficiency	Common carp
	Vitamin A – deficiency	Jian carp
	Thiamine – deficiency	Barramundi
	Riboflavin – deficiency	Barramundi
	Pantothenic acid – deficiency	Blue tilapia, hybrid striped bass, Malabar grouper
	Vitamin C – deficiency	Giant tiger prawn, largemouth bass
Lethargy	Riboflavin – deficiency	Blue tilapia, rainbow trout
	Pyridoxine – deficiency	Hybrid red tilapia, Indian catfish
	Pantothenic acid – deficiency	Spotted rose snapper
	Nicotinic acid – deficiency	Indian catfish
	Biotin – deficiency	Nile tilapia
	Vitamin C – deficiency	Atlantic salmon, rainbow trout
Tetany	Phosphorus – deficiency	Sunshine bass
	Potassium – deficiency	Chinook salmon
	Vitamin D_3 – deficiency	Rainbow trout
	Pyridoxine – deficiency	Indian catfish
Aggressive behavior & cannibalism	Tryptophan – deficiency	Atlantic cod, Barramundi, Brycon, Grouper, pikeperch

Table 8.1 (Continued)

Major pathology sign	Nutrient/ANF/ contaminant	Species
Blood pathology: anemia	Vitamin A – deficiency	Cherry salmon
	Vitamin E – deficiency	Atlantic salmon, cherry/amago salmon
	Vitamin K_3 – deficiency	Fish (general)
	Riboflavin – deficiency	Blue tilapia, brook trout
	Pantothenic acid – deficiency	Blue tilapia, brook trout, hybrid striped bass, Malabar grouper, spotted snakehead
	Nicotinic acid – deficiency	Indian catfish
	Vitamin C – deficiency	African catfish, Atlantic salmon, spotted snakehead (hypochromic anemia)
	Oxidized lipid	Channel catfish, hybrid catfish (hemolytic anemia), rainbow trout (microcytic anemia)
	Zearalenone toxicity	European seabass (macrocytic hypochromic anemia)
Reduced hemoglobin	Lysine – deficiency	Black seabream, Silver perch
	Taurine – deficiency	Yellow catfish
	Threonine – deficiency	Blunt snout bream
	EFA – deficiency	Gilthead seabream
	Phosphorus – deficiency	Stinging catfish, tambaqui
	Iron – deficiency	Atlantic salmon, channel catfish
	Zinc – deficiency	Siberian sturgeon
	Copper – deficiency	Spotted snakehead
	Vitamin A – deficiency	Wuchang bream
	Vitamin D_3 – deficiency	Hybrid tilapia
	Vitamin E – deficiency	Rainbow trout, Siberian sturgeon, spotted snakehead
	Thiamine – deficiency	Indian major carp
	Pyridoxine – deficiency	Indian major carp, orange-spotted grouper
	Pantothenic acid – deficiency	Spotted snakehead
	Nicotinic acid – deficiency	Spotted snakehead
	Folic acid – deficiency	Nile tilapia
	Biotin – deficiency	Catla
	Vitamin C – deficiency	African catfish, rainbow trout
	Oxidized oil	Hybrid catfish, rainbow trout
	Gossypol toxicity	Channel catfish
	Aflatoxin	Nile tilapia, Rohu
	Deoxynivalenol	Rainbow trout
	Fumonisin	Channel catfish
	Zearalenone	European seabass

Table 8.1 (Continued)

Major pathology sign	Nutrient/ANF/ contaminant	Species
Reduced hematocrit	Lysine – deficiency	Cobia, Silver perch, Yellowtail
	Taurine – deficiency	Yellow catfish
	Phosphorus – deficiency	Stinging catfish, tambaqui
	Iron – deficiency	Atlantic salmon, channel catfish
	Zinc – deficiency	Siberian sturgeon
	Copper – deficiency	Spotted snakehead
	Vitamin E – deficiency	Hybrid striped bass, rainbow trout, Siberian sturgeon, spotted snakehead
	Thiamine – deficiency	Indian major carp
	Riboflavin – deficiency	Hybrid red tilapia
	Pyridoxine – deficiency	Hybrid red tilapia, Indian major carp
	Pantothenic acid – deficiency	Spotted snakehead
	Nicotinic acid – deficiency	Spotted snakehead, Indian catfish
	Folic acid – deficiency	Channel catfish, Nile tilapia
	Biotin – deficiency	Catla
	Vitamin C – deficiency	African catfish, hybrid tilapia
	Oxidized oil	Rainbow trout, yellowtail
	Gossypol toxicity	Channel catfish
	Aflatoxin	Channel catfish, Nile tilapia, Rohu
	Fumonisin	African catfish, channel catfish, Nile tilapia
	Deoxynivalenol	Atlantic salmon
	Trichothecene	Channel catfish
	Zearalenone	European seabass
Eye pathology: cataract	Histidine – deficiency	Atlantic salmon
	Methionine – deficiency	Arctic charr
	Tryptophan – deficiency	Chum salmon
	Zinc – deficiency	Rainbow trout
	Riboflavin – deficiency	Barramundi, blue/red tilapia, rainbow trout, sunshine bass
	Vitamin C – deficiency	Yellow perch
	Oxidized lipid	Nile tilapia
Eye pathology: exophthalmia	Vitamin A – deficiency	Jian carp
	Vitamin C – deficiency	Ayu, largemouth bass, rainbow trout
	Oxidized lipid	Nile tilapia

Table 8.1 (Continued)

Major pathology sign	Nutrient/ANF/ contaminant	Species
Fin/skin erosion	Lysine – deficiency	Rainbow trout
	Tryptophan – deficiency	Giant tiger prawn
	EFA – deficiency	Rainbow trout, red drum
	Zinc – deficiency	Common carp, rainbow trout
	Vitamin A – deficiency	Greasy grouper
	Riboflavin – deficiency	Blue tilapia, rainbow trout
	Pyridoxine – deficiency	Hybrid red tilapia
	Pantothenic acid – deficiency	Spotted rose snapper
	Vitamin C – deficiency	Channel catfish, spotted rose snapper, Vundu catfish
Gill structure alterations	Histidine – deficiency	Grass carp
	Valine – deficiency	Grass carp
	EFA – deficiency	Turbot
	Pantothenic acid – deficiency	Barramundi, channel catfish, hybrid striped bass, lake trout, Malabar grouper
	Vitamin C – deficiency	Ayu, channel catfish, giant tiger prawn, Pacu, rainbow trout, walleye
Granulomatosis:	EFA – deficiency	Gilthead seabream (liver granuloma)
	Vitamin C – deficiency	Gilthead seabream, turbot
Heart pathology:	EFA deficiency	Atlantic salmon (increased heart weight & atheriosclerotic alterations), rainbow trout (heart myopathy)
	Copper – deficiency	Whiteleg shrimp (enlarged heart)
	Vitamin E – deficiency	Common carp (enlarged heart), red drum (heart edema)
Hemorrhage pathology:	EFA – deficiency	Gilthead seabream (internal organs)
	Vitamin A – deficiency	Greasy grouper (skin), Jian carp (skin, fins & mouth)
	Thiamine – deficiency	Japanese eel
	Riboflavin – deficiency	Common carp (skin & fins)
	Pyridoxine – deficiency	Indian catfish (intestine & kidney)
	Pantothenic acid – deficiency	Blue tilapia
	Nicotinic acid – deficiency	Indian catfish, spotted snakehead (skin)
	Vitamin C – deficiency	African catfish (spinal column), gilthead seabream, rainbow trout, yellow perch
	Oxidized lipids	Nile tilapia
	Aflatoxin toxicity	European seabass

Table 8.1 (Continued)

Major pathology sign	Nutrient/ANF/ contaminant	Species
Liver pathology: fatty liver	EFA – deficiency	Nile tilapia
	Vitamin D_3 – deficiency	Rainbow trout
	Choline – deficiency	Atlantic salmon, blunt snout bream, white sturgeon
	Oxidized lipids	Channel catfish, grass carp
Liver lipoid degeneration LLD	Oxidized lipids	Atlantic salmon, rainbow trout
Liver/body ceroid accumulation	Vitamin E – deficiency	Blue tilapia, channel catfish, common carp
	Vitamin C – deficiency	African catfish
	Oxidized lipids	Atlantic salmon, hybrid catfish, Nile tilapia
Liver pathology: steatitis	Oxidized lipids	Nile tilapia
Liver pathology: green liver	Taurine – deficiency	Red sea bream, totoaba, yellowtail
Muscular dystrophy	Vitamin E – deficiency	Atlantic salmon, Barramundi, common carp, rainbow trout
	Oxidized lipids	Channel catfish, common carp, turbot
Pigmentation: dark color	Vitamin E – deficiency	Hybrid striped bass
	Thiamine – deficiency	Barramundi, channel catfish, rainbow trout
	Riboflavin – deficiency	Barramundi, sunshine bass
	Pantothenic acid – deficiency	Barramundi
	Biotin – deficiency	Indian catfish, Japanese seabass, walking catfish
	Vitamin C – deficiency	Channel catfish, largemouth bass, spotted rose snapper
Depigmentation	Vitamin E – deficiency	Atlantic salmon
	Vitamin C – deficiency	Gilthead seabream
	Oxidized lipids	Channel catfish, Chinese longsnout catfish
Shock syndrome	EFA – deficiency	Greasy grouper, palmetto bass, rainbow trout, red drum
	Thiamine – deficiency	Barramundi
Short-body dwarfism	Calcium – deficiency	Nile tilapia
	Manganese – deficiency	Rainbow trout
	Riboflavin – deficiency	Blue tilapia, channel catfish, hybrid red tilapia

Table 8.1 (Continued)

Major pathology sign	Nutrient/ANF/ contaminant	Species
Skeletal pathology: lordosis	EFA – deficiency	Grass carp
	Vitamin D_3 – deficiency	Rainbow trout (lordosis-like droopy tail sundrome)
	Vitamin E – deficiency	Common carp
	Vitamin C – deficiency	Atlantic halibut, Atlantic salmon, channel catfish, Coho salmon, largemouth bass, rainbow trout, red drum, spotted snakehead, Vundu catfish, walleye
	Oxidized lipids	Atlantic halibut, Nile tilapia
Skeletal pathology: scoliosis	Tryptophan – deficiency	Chum salmon
	Phosphorus – deficiency	Striped bass
	Vitamin C – deficiency	Atlantic halibut, Atlantic salmon, channel catfish, Coho salmon, rainbow trout, red drum, spotted snakehead, yellow perch
	Oxidized lipid	Atlantic halibut
	Vitamin A – toxicity	Atlantic halibut

antinutrients and mycotoxins, are their direct negative effects on fish/shrimp disease resistance and immune function. For example, nutrient deficiencies which have been reported to have a negative effect on disease resistance and immune function, include (1) essential amino acids – arginine, isoleucine, methionine, phenylalanine, taurine, threonine, tryptophan, and valine, (2) EFA deficiency, (3) trace minerals (copper, trivalent chromium), (4) vitamin A, vitamin E, thiamine, riboflavin, pantothenic acid, nicotinic acid, folic acid, choline, and vitamin C, and (5) or through the direct negative effects of oxidized lipids and mycotoxins on disease resistance and immune function.

References

1. Waagbø, R. (1994). The impact of nutritional factors on the immune system in Atlantic salmon, *Salmo salar* L.: a review *Aquaculture Research*, 25(2):175–197. https://doi.org/10.1111/j.1365–2109.1994.tb00573.x.
2. Ashley, P.J. (2007). Fish welfare: current issues in aquaculture. *Applied Animal Behaviour Science*, 104(3–4):199–235. https://doi.org/10.1016/j.applanim.2006.09.001/.
3. Chen, Y.H. and He, J.G. (2019). Effects of environmental stress on shrimp innate immunity and white spot syndrome virus infection. *Fish and Shellfish Immunology*, 84:744–755. https://doi.org/10.1016/j.fsi.2018.10.069.
4. Davis, K.B. (2006). Management of physiological stress in finfish aquaculture. *North American Journal of Aquaculture*, 68(2):116–121. https://doi.org/10.1577/A05–007.1.
5. Oliva-Teles, A. (2012). Nutrition and health of aquaculture fish. *Journal of Fish Diseases*, 35(2):83–108. https://doi.org/10.1111/j.1365–2761.2011.01333.x.
6. Sung, Y.Y., MacRae, T.H., Sorgeloos, P. and Bossier, P. (2011). Stress response for disease control in aquaculture. *Reviews in Aquaculture*, 3(3):120–137. https://doi.org/10.1111/j.1753–5131.2011.01049.x.
7. Tacon, A.G.J. (1992). *Nutritional fish pathology – morphological signs of nutrient deficiency and toxicity in farmed fish.* FAO Technical Paper No. 330. 75 pp. FAO, Rome.
8. Buentello, J.A. and Gatlin 3rd, D.M. (2001). Effects of elevated dietary arginine on resistance of channel catfish to exposure to Edwardsiella ictaluri. *Journal of Aquatic Animal Health*, 13(3):194–201. https://doi.org/10.1577/1548–8667(2001)013<0194:EOEDAO>2.0.CO;2.

9. Buentello, J.A., Reyes-Becerril, M., de Jesús Romero-Geraldo, M. and de Jesús Ascencio-Valle, F. (2007). Effects of dietary arginine on hematological parameters and innate immune function of channel catfish. *Journal of Aquatic Animal Health*, 19(3):195–203. https://doi.org/10.1577/H07–004.1.
10. Pohlenz, C., Buentello, A., Helland, le J., S. and Gatlin 3rd, D.M. (2014). Effects of dietary arginine supplementation on growth, protein optimization and innate immune response of channel catfish *Ictalurus punctatus* (Rafinesque 1818). *Aquaculture Research*, 45(3):491–500. https://doi.org/10.1111/j.1365–2109.2012.03252.x.
11. Lin, H., Tan, X., Zhou, C., Niu, J., Xia, D., Huang, Z., Wang, J. and Wang, Y. (2015). Effect of dietary arginine levels on the growth performance, feed utilization, non-specific immune response and disease resistance of juvenile golden pompano *Trachinotus ovatus Aquaculture*, 437:382–389. https://doi.org/10.1016/j.aquaculture.2014.12.025.
12. Chen, G., Liu, Y., Jiang, J., Jiang, W., Kuang, S., Tang, L., Tang, W., Zhang, Y.A., Zhou, X. and Feng, L. (2015). Effect of dietary arginine on the immune response and gene expression in head kidney and spleen following infection of Jian carp with Aeromonas hydrophila. *Fish and Shellfish Immunology*, 44(1):195–202. https://doi.org/10.1016/j.fsi.2015.02.027.
13. Wang, B., Feng, L., Chen, G.F., Jiang, W.D., Liu, Y. et al. (2016). Jian carp (*Cyprinus carpio* var. *Jian*) intestinal immune responses, antioxidant status and tight junction protein mRNA expression are modulated via Nrf2 and PKC in response to dietary arginine deficiency. *Fish and Shellfish Immunology*, 51:116–124. https://doi.org/10.1016/j.fsi.2015.10.032.
14. Zhou, H., Chen, N., Qiu, X., Zhao, M. and Jin, L. (2012). Arginine requirement and effect of arginine intake on immunity in largemouth bass, *Micropterus salmoides. Aquaculture Nutrition*, 18(1):107–116. https://doi.org/10.1111/j.1365–2095.2011.00886.x.
15. Vianna, R.A., Chideroli, R.T., da Costa, A.R., Ribeiro Filho, O.P., de Oliveira, L.L., Donzele, J.L., Lanna, E.A.T., Gonçalves, D.D. and Pereira, U.D.P. (2020). Effect of experimental arginine supplementation on the growth, immunity and resistance of tilapia fingerlings to *Streptococcus agalactiae. Aquaculture Research*, 51(3):1276–1283. https://doi.org/10.1111/are.14478.
16. Yue, Y., Zou, Z., Zhu, J., Li, D., Xiao, W., Han, J. and Yang, H. (2015). Effects of dietary arginine on growth performance, feed

utilization, haematological parameters and non-specific immune responses of juvenile Nile tilapia (*Oreochromis niloticus* L.). *Aquaculture Research*, 46(8):1801–1809. https://doi.org/10.1111/are.12333.

17. Cheng, Z., Buentello, A. and Gatlin 3rd, D.M. (2011). Effects of dietary arginine and glutamine on growth performance, immune responses and intestinal structure of red drum, *Sciaenops ocellatus*. *Aquaculture*, 319(1–2):247–252. https://doi.org/10.1016/j.aquaculture.2011.06.025.
18. Rahimnejad, S. and Lee, K.J. (2014a). Dietary arginine requirement of juvenile red sea bream *Pagrus major*. *Aquaculture*, 434:418–424. https://doi.org/10.1016/j.aquaculture.2014.09.003.
19. Costas, B., Conceição, L.E., Dias, J., Novoa, B., Figueras, A. and Afonso, A. (2011). Dietary arginine and repeated handling increase disease resistance and modulate innate immune mechanisms of Senegalese sole (*Solea senegalensis* Kaup, 1858). *Fish and Shellfish Immunology*, 31(6):838–847. https://doi.org/10.1016/j.fsi.2011.07 024.
20. Cheng, Z., Gatlin 3rd, D.M. and Buentello, A. (2012). Dietary supplementation of arginine and/or glutamine influences growth performance, immune responses and intestinal morphology of hybrid striped bass (*Morone chrysops* × *Morone saxatilis*). *Aquaculture*, 362:39–43.
21. Costas, B., Rego, P.C.N.P., Conceicao, L.E.C., Dias, J. and Afonso, A. (2013). Dietary arginine supplementation decreases plasma cortisol levels and modulates immune mechanisms in chronically stressed turbot (*Scophthalmus maximus*). *Aquaculture Nutrition*, 19: 25–38.
22. Zhang, K., Mai, K., Xu, W., Liufu, Z., Zhang, Y., Peng, M., Chen, J. and Ai, Q. (2017). Effects of dietary arginine and glutamine on growth performance, nonspecific immunity, and disease resistance in relation to arginine catabolism in juvenile turbot (*Scophthalmus maximus* L.). *Aquaculture*, 468:246–254. https://doi.org/10.1016/j.aquaculture.2016.10.021.
23. Wang, L. (2017). Effects of dietary arginine levels on intestinal digestive enzyme activity and morphology in juvenile hybrid sturgeon. *Dalian Hai Yang da Xue Xue Bao, Dalian*, 32(1), pp. 51–55.
24. Zhou, Q., Jin, M., Elmada, Z.C., Liang, X. and Mai, K. (2015). Growth, immune response and resistance to *Aeromonas hydrophila* of

juvenile yellow catfish, *Pelteobagrus fulvidraco*, fed diets with different arginine levels. *Aquaculture*, 437:84–91. https://doi.org/10.1016/j.aquaculture.2014.11.030.

25. Zhao, J., Liu, Y., Jiang, J., Wu, P., Chen, G., Jiang, W., Li, S., Tang, L., Kuang, S., Feng, L. and Zhou, X. (2012). Effects of dietary isoleucine on growth, the digestion and absorption capacity and gene expression in hepatopancreas and intestine of juvenile Jian carp (*Cyprinus carpio* var. *Jian*). *Aquaculture*, 368:117–128. https://doi.org/10.1016/j.aquaculture.2012.09.019.
26. Zhao, J., Feng, L., Liu, Y., Jiang, W., Wu, P., Jiang, J., Zhang, Y. and Zhou, X. (2014). Effect of dietary isoleucine on the immunity, antioxidant status, tight junctions and microflora in the intestine of juvenile Jian carp (*Cyprinus carpio* var. *Jian*). *Fish and Shellfish Immunology*, 41(2):663–673. https://doi.org/10.1016/j.fsi.2014.10.002.
27. Rahimnejad, S. and Lee, K.J. (2014b). Dietary isoleucine influences non-specific immune response in juvenile olive flounder (*Paralichthys olivaceus*). *Turkish Journal of Fisheries and Aquatic Sciences*, 14(4):853–862.
28. Kwasek, K., Dabrowski, K., Nynca, J., Wojno, M. and Wick, M. (2014). The influence of dietary lysine on Yellow Perch maturation and the quality of sperm. *North American Journal of Aquaculture*, 76(2):119–126. https://doi.org/10.1080/15222055.2013.856826.
29. Yang, S.D., Liu, F.G. and Liou, C.H. (2011). Assessment of dietary lysine requirement for silver perch (*Bidyanus bidyanus*) juveniles. *Aquaculture*, 312(1–4):102–108. https://doi.org/10.1016/j.aquaculture.2010.12.011.
30. Zhou, F., Shao, J., Xu, R., Ma, J. and Xu, Z. (2010). Quantitative L-lysine requirement of juvenile black sea bream (*Sparus macrocephalus*). *Aquaculture Nutrition*, 16(2):194–204. https://doi.org/10.1111/j.1365–2095.2009.00651.x.
31. Zhou, Q.C., Wu, Z.H., Chi, S.Y. and Yang, Q.H. (2007). Dietary lysine requirement of juvenile cobia (*Rachycentron canadum*). *Aquaculture*, 273(4):634–640. https://doi.org/10.1016/j.aquaculture.2007.08.056.
32. Ruchimat, T., Masumoto, T., Hosokawa, H., Itoh, Y. and Shimeno, S. (1997). Quantitative lysine requirement of yellowtail (*Seriola quinqueradiata*). *Aquaculture*, 158(3–4):331–339. https://doi.org/10.1016/S0044–8486(97)00215–9.

33. Anderson, J.S., Lall, S.P., Anderson, D.M. and Mcniven, M.A. (1993). Quantitative dietary lysine requirement of Atlantic salmon (*Salmo salar*) fingerlings. *Canadian Journal of Fisheries and Aquatic Sciences*, 50(2):316–322. https://doi.org/10.1139/f93–037.
34. Borlongan, I.G. and Benitez, L.V. (1990). Quantitative lysine requirement of milkfish (*Chanos chanos*) juveniles. *Aquaculture*, 87 (3–4):341–347. https://doi.org/10.1016/0044–8486(90)90071-T.
35. Ketola, H.G. (1983). Requirement for dietary lysine and arginine by fry of rainbow trout. *Journal of Animal Science*, 56(1):101–107. https://doi.org/10.2527/jas1983.561101x.
36. He, J.Y., Long, W.Q., Han, B., Tian, L.X., Yang, H.J., Zeng, S.L. and Liu, Y.J. (2017). Effect of dietary l-methionine concentrations on growth performance, serum immune and antioxidative responses of juvenile Nile tilapia, *Oreochromis niloticus*. *Aquaculture Research*, 48(2):665–674.
37. Elmada, C.Z., Huang, W., Jin, M., Liang, X., Mai, K. and Zhou, Q. (2016). The effect of dietary methionine on growth, antioxidant capacity, innate immune response and disease resistance of juvenile yellow catfish (*Pelteobagrus fulvidraco*). *Aquaculture Nutrition*, 22(6):1163–1173. https://doi.org/10.1111/anu.12363.
38. Machado, M., Azeredo, R., Díaz-Rosales, P., Afonso, A., Peres, H., Oliva-Teles, A. and Costas, B. (2015). Dietary tryptophan and methionine as modulators of European seabass (*Dicentrarchus labrax*) immune status and inflammatory response. *Fish and Shellfish Immunology*, 42(2):353–362. https://doi.org/10.1016/j.fsi.2014.11.024.
39. Tang, L., Wang, G.X., Jiang, J., Feng, L., Yang, L., Li, S.H., Kuang, S.Y. and Zhou, X.Q. (2009). Effect of methionine on intestinal enzymes activities, microflora and humoral immune of juvenile Jian carp (*Cyprinus carpio* var. *Jian*). *Aquaculture Nutrition*, 15(5): 477–483. https://doi.org/10.1111/j.1365–2095.2008.00613.x.
40. Simmons, L., Moccia, R.D., Bureau, D.P., Sivak, J.G. and Herbert, K. (1999). Dietary methionine requirement of juvenile Arctic charr *Salvelinus alpinus* (L.). *Aquaculture Nutrition*, 5(2):93–100. https://doi.org/10.1046/j.1365–2095.1999.00097.x.
41. Rollin, X., Mambrini, M., Abboudi, T., Larondelle, Y. and Kaushik, S.J. (2003). The optimum dietary indispensable amino acid pattern for growing Atlantic salmon (*Salmo salar* L.) fry. *British Journal of Nutrition*, 90(5):865–876. https://doi.org/10.1079/BJN2003973.

42. Li, W., Feng, L., Liu, Y., Jiang, W.D., Kuang, S.Y., Jiang, J., Li, S.H., Tang, L. and Zhou, X.Q. (2015). Effects of dietary phenylalanine on growth, digestive and brush border enzyme activities and antioxidant capacity in the hepatopancreas and intestine of young grass carp (*Ctenopharyngodon idella*). *Aquaculture Nutrition*, 21(6):913–925. https://doi.org/10.1111/anu.12223.
43. Feng, L., Li, W., Liu, Y., Jiang, W.D., Kuang, S.Y., Jiang, J., Tang, L., Wu, P., Tang, W.N., Zhang, Y.A. and Zhou, X.Q. (2015). Dietary phenylalanine-improved intestinal barrier health in young grass carp (*Ctenopharyngodon idella*) is associated with increased immune status and regulated gene expression of cytokines, tight junction proteins, antioxidant enzymes and related signalling molecules. *Fish & Shellfish immunology*, 45(2):495–509. https://doi.org/10.1016/j.fsi.2015.05.001.
44. Habte-Tsion, H.M., Ren, M., Liu, B., Ge, X., Xie, J. and Chen, R. (2016). Threonine modulates immune response, antioxidant status and gene expressions of antioxidant enzymes and antioxidant-immune-cytokine-related signaling molecules in juvenile blunt snout bream (*Megalobrama amblycephala*). *Fish and Shellfish Immunology*, 51:189–199. https://doi.org/10.1016/j.fsi.2015.11.033.
45. Habte-Tsion, H.M., Ren, M., Liu, B., Ge, X., Xie, J., Chen, R., Zhou, Q. and Pan, L. (2015). Threonine influences the absorption capacity and brush-border enzyme gene expression in the intestine of juvenile blunt snout bream (*Megalobrama amblycephala*). *Aquaculture*, 448:436–444. https://doi.org/10.1016/j.aquaculture.2015.06.030.
46. Habte-Tsion, H.M., Ge, X., Liu, B., Xie, J., Ren, M., Zhou, Q., Miao, L., Pan, L. and Chen, R. (2015). A deficiency or an excess of dietary threonine level affects weight gain, enzyme activity, immune response and immune-related gene expression in juvenile blunt snout bream (*Megalobrama amblycephala*). *Fish and Shellfish Immunology*, 42(2):439–446. https://doi.org/10.1016/j.fsi.2014.11.021.
47. Gao, Y.J., Yang, H.J., Liu, Y.J., Chen, S.J., Guo, D.Q., Yu, Y.Y. and Tian, L.X. (2014). Effects of graded levels of threonine on growth performance, biochemical parameters and intestine morphology of juvenile grass carp *Ctenopharyngodon idella*. *Aquaculture*, 424: 113–119. https://doi.org/10.1016/j.aquaculture.2013.12.043.
48. Hoseini, S.M., Taheri Mirghaed, A. and Ghelichpour, M. (2020). Effects of dietary tryptophan levels and fish stocking density on

immunological and antioxidant responses and bactericidal activity against *Aeromonas hydrophila* in rainbow trout (*Oncorhynchus mykiss*). *Aquaculture Research*, 51(4):1455–1463. https://doi.org/10.1111/are.14491.

49. Zhao, Y., Wu, X.Y., Xu, S.X., Xie, J.Y., Xiang, K.W. et al. (2019). Dietary tryptophan affects growth performance, digestive and absorptive enzyme activities, intestinal antioxidant capacity, and appetite and GH-IGF axis-related gene expression of hybrid catfish (*Pelteobagrus vachelli*♀ × *Leiocassis longirostris*♂). *Fish Physiology and Biochemistry*, 45(5):1627–1647. https://doi.org/10.1007/s10695–019–00651–4.
50. Cabanillas-Gámez, M., López, L.M., Galaviz, M.A., True, C.D. and Bardullas, U. (2018). Effect of L-tryptophan supplemented diets on serotonergic system and plasma cortisol in *Totoaba macdonaldi* (Gilbert, 1890) juvenile exposed to acute stress by handling and hypoxia. *Aquaculture Research*, 49(2):847–857. https://doi.org/10.1111/are.13529.
51. Kumar, P., Kailasam, M., Sethil, S.N., Sukumaran, K., Biswas, G., Subburaj, R., Thiagarajan, S., Ghoshal, T.K. and Vijayan, K.K. (2017). Effect of dietary L-tryptophan on cannibalism, growth and survival of Asian seabass, *Lates calcarifer* (Bloch, 1790) fry. *Indian Journal of Fisheries, (New Delhi)*, 64(2):28–32. https://doi.org/10.21077/ijf.2017.64.2.61333–05.
52. Hoseini, S.M., Mirghaed, A.T., Mazandarani, M. and Zoheiri, F. (2016). Serum cortisol, glucose, thyroid hormones' and non-specific immune responses of Persian sturgeon, *Acipenser persicus* to exogenous tryptophan and acute stress. *Aquaculture*, 462:17–23. https://doi.org/10.1016/j.aquaculture.2016.04.031.
53. Król, J. and Zakęś, Z. (2016). Effect of dietary L-tryptophan on cannibalism, survival and growth in pikeperch *Sander lucioperca* (L.) post-larvae. *Aquaculture International*, 24(2):441–451. https://doi.org/10.1007/s10499–015–9936–1.
54. Sun, Y., Qiu, J. and Wang, G. (2016). Effects of tryptophan supplemented in low protein diets on apparent digestibility coefficients, digestive enzyme activity and amino acids composition of *Litopenaeus vannamei*. *Acta Hydrobiologica Sinica*/Shuisheng Shengwu Xuebao, 40(4):720–727.
55. Jiang, W.D., Wen, H.L., Liu, Y., Jiang, J., Kuang, S.Y. et al. (2015). The tight junction protein transcript abundance changes and oxidative damage by tryptophan deficiency or excess are related to the

modulation of the signalling molecules, NF-κB p65, TOR, caspase-(3, 8, 9) and Nrf2 mRNA levels, in the gill of young grass carp (*Ctenopharyngodon idellus*). *Fish and Shellfish Immunology*, 46(2):168–180. https://doi.org/10.1016/j.fsi.2015.06.002.

56. Machado, M., Azeredo, R., Díaz-Rosales, P., Afonso, A., Peres, H., Oliva-Teles, A. and Costas, B. (2015). Dietary tryptophan and methionine as modulators of European seabass (*Dicentrarchus labrax*) immune status and inflammatory response. *Fish and Shellfish Immunology*, 42(2):353–362. https://doi.org/10.1016/j.fsi.2014.11.024.
57. Wolkers, C.P.B., Serra, M., Szawka, R.E. and Urbinati, E.C. (2014). The time course of aggressive behaviour in juvenile matrinxã *Brycon amazonicus* fed with dietary L-tryptophan supplementation. *Journal of Fish Biology*, 84(1):45–57. https://doi.org/10.1111/jfb.12252.
58. Tang, L., Feng, L., Sun, C.Y., Chen, G.F., Jiang, W.D., Hu, K., Liu, Y., Jiang, J., Li, S.H., Kuang, S.Y. and Zhou, X.Q. (2013). Effect of tryptophan on growth, intestinal enzyme activities and TOR gene expression in juvenile Jian carp (*Cyprinus carpio* var. *Jian*): Studies in vivo and in vitro. *Aquaculture*, 412:23–33. https://doi.org/10.1016/j.aquaculture.2013.07.002.
59. Basic, D., Schjolden, J., Krogdahl, Å., von Krogh, K., Hillestad, M., Winberg, S., Mayer, I., Skjerve, E. and Höglund, E. (2013). Changes in regional brain monoaminergic activity and temporary down-regulation in stress response from dietary supplementation with L-tryptophan in Atlantic cod (*Gadus morhua*). *British Journal of Nutrition*, 109(12):2166–2174. https://doi.org/10.1017/S0007114512004345.
60. Hoseini, S.M. and Hosseini, S.A. (2010). Effect of dietary L-tryptophan on osmotic stress tolerance in common carp, *Cyprinus carpio*, juveniles. *Fish Physiology and Biochemistry*, 36(4):1061–1067. https://doi.org/10.1007/s10695–010–9383-x.
61. Tejpal, C.S., Pal, A.K., Sahu, N.P., Kumar, J.A., Muthappa, N.A., Vidya, S. and Rajan, M.G. (2009). Dietary supplementation of L-tryptophan mitigates crowding stress and augments the growth in *Cirrhinus mrigala* fingerlings. *Aquaculture*, 293(3–4):272–277. https://doi.org/10.1016/j.aquaculture.2008.09.014.
62. Höglund, E., Bakke, M.J., Øverli, Ø., Winberg, S. and Nilsson, G.E. (2005). Suppression of aggressive behaviour in juvenile Atlantic cod (*Gadus morhua*) by l-tryptophan

supplementation. *Aquaculture*, 249(1–4):525–531. https://doi.org/10.1016/j.aquaculture.2005.04.028.

63. Hseu, J.R., Lu, F.I., Su, H.M., Wang, L.S., Tsai, C.L. and Hwang, P.P. (2003). Effect of exogenous tryptophan on cannibalism, survival and growth in juvenile grouper, *Epinephelus coioides*. *Aquaculture*, 218(1–4):251–263. https://doi.org/10.1016/S0044–8486(02)00503–3.
64. Lepage, O., Tottmar, O. and Winberg, S. (2002). Elevated dietary intake of L-tryptophan counteracts the stress-induced elevation of plasma cortisol in rainbow trout (*Oncorhynchus mykiss*). *Journal of Experimental Biology*, 205(23):3679–3687. https://doi.org/10.1242/jeb.205.23.3679.
65. Akiyama, T., Shiraishi, M., Yamamoto, T. and Unuma, T. (1996). Effect of dietary tryptophan on maturation of Ayu *Plecoglossus altivelis*. *Fisheries Science*, 62(5):776–782. https://doi.org/10.2331/fishsci.62.776.
66. Han, A., Liang, Y., Gao, C. and Sun, M. (1995). Requirement for dietary arginine, methionine, phenylalanine and tryptophan by juvenile of the shrimp *Penaeus monodon*. *Journal of Fishery Sciences of China*, 2(2):7–14.
67. Akiyama, T., Mori, K. and Murai, T. (1986). Effects of temperature on the incidence of scoliosis and cataract in chum salmon fry caused by tryptophan deficiency. *Nippon Suisan Gakkaishi*, 52(11):2039.
68. Feng, L., Luo, J.B., Jiang, W.D., Liu, Y., Wu, P., Jiang, J., Kuang, S.Y., Tang, L., Zhang, Y.A. and Zhou, X.Q. (2015). Changes in barrier health status of the gill for grass carp (*Ctenopharyngodon idella*) during valine deficiency: regulation of tight junction protein transcript, antioxidant status and apoptosis-related gene expression. *Fish and Shellfish Immunology*, 45(2):239–249. https://doi.org/10.1016/j.fsi.2015.04.023.
69. Rahimnejad, S. and Lee, K.J. (2013). Dietary valine requirement of juvenile red sea bream *Pagrus major*. *Aquaculture*, 416:212–218. https://doi.org/10.1016/j.aquaculture.2013.09.026.
70. Dong, M., Feng, L., Kuang, S.Y., Liu, Y., Jiang, J., Hu, K., Jiang, W.D., Li, S.H., Tang, L. and Zhou, X.Q. (2013). Growth, body composition, intestinal enzyme activities and microflora of juvenile Jian carp (*Cyprinus carpio* var. *Jian*) fed graded levels of dietary valine. *Aquaculture Nutrition*, 19(1):1–14. https://doi.org/10.1111/j.1365–2095.2011.00926.x.

71. Saleh, N.E., Wassef, E.A. and Ashry, A.M. (2020). Is a taurine supplement necessary in fishmeal-based feeds for juvenile European sea bass (*Dicentrarchus labrax*)? *Aquaculture International*, 28(1): 321–333. https://doi.org/10.1007/s10499–019–00464–5.
72. Tan, X., Sun, Z., Zhu, X. and Ye, C. (2018). Dietary supplementation with taurine improves ability to resist ammonia stress in hybrid snakehead (*Channa maculatus*♀ × *Channa argus*♂). *Aquaculture Research*, 49(10):3400–3410. https://doi.org/10.1111/are.13804.
73. Han, Y., Koshio, S., Jiang, Z., Ren, T., Ishikawa, M., Yokoyama, S. and Gao, J. (2014). Interactive effects of dietary taurine and glutamine on growth performance, blood parameters and oxidative status of Japanese flounder *Paralichthys olivaceus. Aquaculture*, 434:348–354. https://doi.org/10.1016/j.aquaculture.2014.08.036.
74. Lu, Y., You, C., Wang, S. and Li, Y. (2014). Physiological changes in *Siganus canaliculatus* after shallow water stress and the anti-stress effects of taurine. *Acta Hydrobiologica Sinica*/Shuisheng Shengwu Xuebao, 38(1):68–74.
75. Kato, K., Yamamoto, M., Peerapon, K., Fukada, H., Biswas, A. et al. (2014). Effects of dietary taurine levels on epidermal thickness and scale loss in red sea bream, *Pagrus major. Aquaculture Research*, 45(11):1818–1824. https://doi.org/10.1111/are.12130.
76. Takagi, S., Murata, H., Goto, T., Ichiki, T., Endo, M., Hatate, H., Yoshida, T., Sakai, T., Yamashita, H. and Ukawa, M. (2006). Efficacy of taurine supplementation for preventing green liver syndrome and improving growth performance in yearling red sea bream *Pagrus major* fed low-fishmeal diet. *Fisheries Science*, 72(6): 1191–1199. https://doi.org/10.1111/j.1444–2906.2006.01276.x.
77. Takagi, S., Murata, H., Goto, T., Hatate, H., Endo, M., Yamashita, H., Miyatake, H. and Ukawa, M. (2010a). Necessity of dietary taurine supplementation for preventing green liver symptom and improving growth performance in yearling red sea bream *Pagrus major* fed non-fishmeal diets based on soy protein concentrate. *Fisheries Science*, 76(1):119. https://doi.org/10.1007/s12562–009–0185-y.
78. Takagi, S., Murata, H., Goto, T., Hatate, H., Endo, M. et al. (2010b). Role of taurine deficiency in inducing green liver symptom and effect of dietary taurine supplementation in improving growth in juvenile red sea bream *Pagrus major* fed non-fishmeal diets based on soy protein concentrate. *Fisheries Science*, 74(6):235–244. https://doi.org/10.1007/s12562–011–0322–2.

79. Zhao, M., Liang, M., Zheng, K., Xiao, D. and Li, Q. (2015). Effects of dietary taurine on the reproductive performance and the larval quality in tonguesole *Cynoglossus semilaevis*. Progress in fishery science. *Yuye Kexue Jinzhan*, 36(3):101–108.
80. Satriyo, T.B., Galaviz, M.A., Salze, G. and Lopez, L.M. (2017). Assessment of dietary taurine essentiality on the physiological state of juvenile *Totoaba macdonaldi*. *Aquaculture Research*, 48(11): 5677–5689. https://doi.org/10.1111/are.13391.
81. Liu, Y., Yang, P., Hu, H., Li, Y., Dai, J., Zhang, Y., Ai, Q., Xu, W., Zhang, W. and Mai, K. (2018). The tolerance and safety assessment of taurine as additive in a marine carnivorous fish, *Scophthalmus maximus* L. *Aquaculture Nutrition*, 24(1):461–471. https://doi.org/10.1111/anu.12579.
82. Li, M., Lai, H., Li, Q., Gong, S. and Wang, R. (2016). Effects of dietary taurine on growth, immunity and hyperammonemia in juvenile yellow catfish *Pelteobagrus fulvidraco* fed all-plant protein diets. *Aquaculture*, 450:349–355. https://doi.org/10.1016/j.aquaculture.2015.08.013.
83. Matsumoto, Y., La Xuan, T., Morioka, K., Fukada, H. and Masumoto, T. (2017). Effects of graded levels of taurine supplement to the no-fishmeal diet on the growth and physiology in juvenile yellowtail *Seriola quinqueradiata*. *Suisanzoshoku = Aquaculture Science*, 65(3):239–246.
84. Yang, Q., Liang, H., Mokrani, A., Ji, K., Yu, H., Ge, X., Ren, M., Xie, J., Pan, L. and Sun, A. (2019). Dietary histidine affects intestinal antioxidant enzyme activities, antioxidant gene expressions and inflammatory factors in juvenile blunt snout bream (*Megalobrama amblycephala*). *Aquaculture Nutrition*, 25(1):249–259. https://doi.org/10.1111/anu.12847.
85. Sambraus, F., Fjelldal, P.G., Remø, S.C., Hevrøy, E.M., Nilsen, T.O., Thorsen, A., Hansen, T.J. and Waagbø, R. (2017). Water temperature and dietary histidine affect cataract formation in Atlantic salmon (*Salmo salar* L.) diploid and triploid yearling smolt. *Journal of Fish Diseases*, 40(9):1195–1212. https://doi.org/10.1111/jfd.12594.
86. Jiang, W.D., Feng, L., Qu, B., Wu, P., Kuang, S.Y., Jiang, J., Tang, L., Tang, W.N., Zhang, Y.A., Zhou, X.Q. and Liu, Y. (2016). Changes in integrity of the gill during histidine deficiency or excess due to depression of cellular anti-oxidative ability, induction of apoptosis,

inflammation and impair of cell-cell tight junctions related to Nrf2, TOR and NF-κB signaling in fish. *Fish and Shellfish Immunology*, 56:111–122. https://doi.org/10.1016/j.fsi.2016.07.002.

87. Gao, Y.J., Liu, Y.J., Chen, X.Q., Yang, H.J., Li, X.F. and Tian, L.X. (2016). Effects of graded levels of histidine on growth performance, digested enzymes activities, erythrocyte osmotic fragility and hypoxia-tolerance of juvenile grass carp *Ctenopharyngodon idella*. *Aquaculture*, 452:388–394. https://doi.org/10.1016/j.aquaculture.2015.11.019.
88. Taylor, J.F., Waagbø, R., Diez-Padrisa, M., Campbell, P., Walton, J., Hunter, D., Matthew, C. and Migaud, H. (2015). Adult triploid Atlantic salmon (*Salmo salar*) have higher dietary histidine requirements to prevent cataract development in seawater. *Aquaculture Nutrition*, 21(1):18–32. https://doi.org/10.1111/anu.12130.
89. Feng, L., Zhao, B., Chen, G., Jiang, W., Liu, Y., Jiang, J., Hu, K., Li, S. and Zhou, X. (2013). Effects of dietary histidine on antioxidant capacity in juvenile Jian carp (*Cyprinus carpio* var. *Jian*). *Fish Physiology and Biochemistry*, 39(3):559–571. https://doi.org/10.1007/s10695–012–9719–9.
90. Zhao, B., Feng, L., Liu, Y., Kuang, S.Y., Tang, L., Jiang, J., Hu, K., Jiang, W.D., Li, S.H. and Zhou, X.Q. (2012). Effects of dietary histidine levels on growth performance, body composition and intestinal enzymes activities of juvenile Jian carp (*Cyprinus carpio* var. *Jian*). *Aquaculture Nutrition*, 18(2):220–232.
91. Waagbø, R., Tröße, C., Koppe, W., Fontanillas, R. and Breck, O. (2010). Dietary histidine supplementation prevents cataract development in adult Atlantic salmon, *Salmo salar* L., in seawater. *British Journal of Nutrition*, 104(10):1460–1470.
92. Breck, O., Bjerkås, E., Campbell, P., Arnesen, P., Haldorsen, P. and Waagbø, R. (2003). Cataract preventative role of mammalian blood meal, histidine, iron and zinc in diets for Atlantic salmon (*Salmo salar* L.) of different strains. *Aquaculture Nutrition*, 9(5):341–350. https://doi.org/10.1046/j.1365–2095.2003.00269.x.
93. Recodo, A.G., Herrera, A.A. and Pascual, F.P. (2000). Histological effects of various dietary L-histidine levels on the midgut gland of *Penaeus monodon Fabricius*. *Asia Life Sciences (Philippines)*, 9(1):53–65.
94. Nunes, A.J., Sá, M.V., Browdy, C.L. and Vazquez-Anon, M. (2014). Practical supplementation of shrimp and fish feeds with crystalline

amino acids. *Aquaculture*, 431:20–27. https://doi.org/10.1016/j.aquaculture.2014.04.003.

95. Li, P., Mai, K., Trushenski, J. and Wu, G. (2009). New developments in fish amino acid nutrition: towards functional and environmentally oriented aquafeeds. *Amino Acids*, 37(1):43–53. https://doi.org/10.1007/s00726–008–0171–1.
96. Wu, G. (2009). Amino acids: metabolism, functions, and nutrition. *Amino Acids*, 37(1):1–17. https://doi.org/10.1007/s00726–009–0269–0.
97. Mao, X., Zeng, X., Qiao, S., Wu, G. and Li, D. (2011). Specific roles of threonine in intestinal mucosal integrity and barrier function. *Frontiers in Bioscience (Elite Edition)*, 3(4):1192–1200. https://doi.org/10.2741/e322.
98. Hoseini, S.M., Pérez-Jiménez, A., Costas, B., Azeredo, R. and Gesto, M. (2019). Physiological roles of tryptophan in teleosts: current knowledge and perspectives for future studies. *Reviews in Aquaculture*, 11(1):3–24. https://doi.org/10.1111/raq.12223.
99. Tacon, A.G.J., Metian, M. and Hasan, M.R. (2009). Feed ingredients and fertilizers for farmed aquatic animals: sources and composition. *FAO Fisheries and aquaculture Technical Paper*. 540. FAO, Rome.
100. Watson, A.M., Barrows, F.T. and Place, A.R. (2015). Leaching of taurine from commercial type aquaculture feeds. *Aquaculture Research*, 46(6):1510–1517. https://doi.org/10.1111/are.12309.
101. Nordgreen, A., Tonheim, S. and Hamre, K. (2009). Protein quality of larval feed with increased concentration of hydrolysed protein: effects of heat treatment and leaching. *Aquaculture Nutrition*, 15(5): 525–536. https://doi.org/10.1111/j.1365–2095.2008.00619.x.
102. Yúfera, M., Kolkovski, S., Fernández-Dıaz, C. and Dabrowski, K. (2002). Free amino acid leaching from a protein-walled microencapsulated diet for fish larvae. *Aquaculture*, 214(1–4):273–287. https://doi.org/10.1016/S0044–8486(01)00902–4.
103. López-Alvarado, J., Langdon, C.J., Teshima, S.I. and Kanazawa, A. (1994). Effects of coating and encapsulation of crystalline amino acids on leaching in larval feeds. *Aquaculture*, 122(4):335–346. https://doi.org/10.1016/0044–8486(94)90342–5.
104. Wei, Y., Xu, H. and Liang, M. (2020). Amino acid absorption and protein synthesis responses of turbot *Scophthalmus maximus* to lysine and leucine in free, dipeptide and tripeptide forms. *Aquaculture Nutrition*, 26(2):358–367. https://doi.org/10.1111/anu.12998.

105. Alam, M.S., Teshima, S., Koshio, S. and Ishikawa, M. (2004). Effects of supplementation of coated crystalline amino acids on growth performance and body composition of juvenile kuruma shrimp *Marsupenaeus japonicus. Aquaculture Nutrition*, 10(5):309–316. https://doi.org/10.1111/j.1365–2095.2004.00316.x.
106. Aragao, C., Colen, R., Ferreira, S., Pinto, W., Conceição, L.E.C. and Dias, J. Pinto, W. & L.E.C. Conceicao et al. (2014). Microencapsulation of taurine in Senegalese sole diets improves its metabolic availability. *Aquaculture*, 431:53–58. https://doi.org/10.1016/j.aquaculture.2014.04.041.
107. Csapó, J., Varga-Visi, E., Lóki, K., Albert, C. and Salamon, S. (2008). The influence of extrusion on loss and racemization of amino acids. *Amino Acids*, 34(2):287–292. https://doi.org/10.1007/s00726–006–0484-x.
108. Velasquez, A., Pohlenz, C., Barrows, F.T., Gaylord, T.G. and Gatlin, D.M. (2015). Assessment of taurine bioavailability in pelleted and extruded diets with red drum *Sciaenops ocellatus. Aquaculture*, 449:2–7. https://doi.org/10.1016/j.aquaculture.2015.03.034.
109. Glencross, B.D. (2009). Exploring the nutritional demand for essential fatty acids by aquaculture species. *Reviews in Aquaculture*, 1(2):71–124. https://doi.org/10.1111/j.1753–5131.2009.01006.x.
110. National Research Council. (2011). *Nutrient requirements of fish and shrimp*. National Academy Press, Washington, DC.
111. Ruyter, B., Røsjø, C., Einen, O. and Thomassen, M.S. (2000). Essential fatty acids in Atlantic salmon: effects of increasing dietary doses of n-6 and n-3 fatty acids on growth, survival and fatty acid composition of liver, blood and carcass. *Aquaculture Nutrition*, 6(2): 119–127. https://doi.org/10.1046/j.1365–2095.2000.00137.x.
112. Seierstad, S.L., Poppe, T.T., Koppang, E.O., Svindland, A., Rosenlund, G., Frøyland, L. and Larsen, S. (2005). Influence of dietary lipid composition on cardiac pathology in farmed Atlantic salmon, *Salmo salar* L. *Journal of Fish Diseases*, 28(11):677–690. https://doi.org/10.1111/j.1365–2761.2005.00671.x.
113. Jin, Y., Angell, I.L., Sandve, S.R., Snipen, L.G., Olsen, Y. and Rudi, K. (2019). Atlantic salmon raised with diets low in long-chain polyunsaturated n-3 fatty acids in freshwater have a mycoplasma-dominated gut microbiota at sea. *Aquaculture Environment Interactions*, 11: 31–39. https://doi.org/10.3354/aei00297.

114. Chim, L., Lemaire, P., Delaporte, M., Le Moullac, G., Galois, R. and Martin, J.L.M. (2001). Could a diet enriched with n-3 highly unsaturated fatty acids be considered a promising way to enhance the immune defences and the resistance of Penaeid prawns to environmental stress? *Aquaculture Research*, 32(2):91–94. https://doi.org/10.1046/j.1365–2109.2001.00534.x.
115. Fukada, H., Taniguchi, E., Morioka, K. and Masumoto, T. (2017). Effects of replacing fish oil with canola oil on the growth performance, fatty acid composition and metabolic enzyme activity of juvenile yellowtail *Seriola quinqueradiata* (Temminck & Schlegel, 1845). *Aquaculture Research*, 48(12):5928–5939. https://doi.org/10.1111/are.13416.
116. Takeuchi, T., Watanabe, T. and Nose, T. (1979). Requirement for essential fatty acids of Chum salmon (*Oncorhynchus keta*) in freshwater environment. *Nihon Suisan Gakkai, Shi; Tokyo*, 45(10): 1319.
117. McKenzie, D.J., Lund, I. and Pedersen, P.B. (2008). Essential fatty acids influence metabolic rate and tolerance of hypoxia in Dover sole (*Solea solea*) larvae and juveniles. *Marine Biology*, 154(6): 1041–1051. https://doi.org/10.1007/s00227–008–0998–8.
118. Mourente, G., Good, J.E., Thompson, K.D. and Bell, J.G. (2007). Effects of partial substitution of dietary fish oil with blends of vegetable oils, on blood leucocyte fatty acid compositions, immune function and histology in European sea bass (*Dicentrarchus labrax* L.). *British Journal of Nutrition*, 98(4):770–779. https://doi.org/10.1017/S000711450773461X.
119. Montero, D., Socorro, J., Tort, L., Caballero, M.J., Robaina, L.E., Vergara, J.M. and Izquierdo, M.S. (2004). Glomerulonephritis and immunosuppression associated with dietary essential fatty acid deficiency in gilthead sea bream, *Sparus aurata* L., juveniles. *Journal of Fish Diseases*, 27(5):297–306. https://doi.org/10.1111/j.1365–2761.2004.00543.x.
120. Rodriguez, C., Cejas, J.R., Martin, M.V., Badia, P., Samper, M. and Lorenzo, A. (1998). Influence of n-3 highly unsaturated fatty acid deficiency on the lipid composition of broodstock gilthead sea bream (*Sparus aurata* L.) and on egg quality. *Fish Physiology and Biochemistry*, 18(2):177–187.
121. Alexi, M.N., Karanikolas, K. and Anastasopoulou, G. (1993). Results of essential fatty acid deficiency on gilthead bream

pathology. *Abstracts of the Panellinio Symposio Okeanografias Kai Alieias.* Perilipseis. Hellenic Symposium on Oceanography and Fisheries, pp. 448–451. National Centre for Marine Research, Athens, Greece. NCMR Association of Employees, Athens, Greece, April 1993.

122. Rodriguez, C., Perez, J.A., Izquierdo, M.S., Mora, J., Lorenzo, A. and Fernandez-Palacios, H. (1994). Essential fatty acid requirements of larval gilthead sea bream, *Sparus aurata* (L.). *Aquaculture Research*, 25(3):295–304. https://doi.org/10.1111/j.1365–2109.1994.tb00692.x.
123. Takeuchi, T., Watanabe, K., Wen-Yue, Y. and Watanabe, T. (1991). Essential fatty acids of grass carp *Ctenopharyngodon Idella.* Nihon Suisan Gakkai, Shi; Tokyo, 57(3):467–473. https://doi.org/10.2331/suisan.57.467.
124. Lim, L.C. (1993). Larviculture of the greasy grouper Epinephelus tauvina F. and the brown-marbled grouper *E. fuscoguttatus* F. in Singapore. *Journal of the World Aquaculture Society*, 24(2):262–274. https://doi.org/10.1111/j.1749–7345.1993.tb00015.x.
125. Xu, H., Ai, Q., Mai, K., Xu, W., Wang, J., Ma, H., Zhang, W., Wang, X. and Liufu, Z. (2010). Effects of dietary arachidonic acid on growth performance, survival, immune response and tissue fatty acid composition of juvenile Japanese seabass, *Lateolabrax japonicus. Aquaculture*, 307(1–2):75–82. https://doi.org/10.1016/j.aquaculture.2010.07.001.
126. Alava, V.R. and Kanazawa, A. (1996). Effect of dietary fatty acids on growth of milkfish *Chanos chanos* fry in brackish water. *Aquaculture*, 144(4):363–369.https://doi.org/10.1016/0044–8486(96)01304-X.
127. Borlongan, I.G. (1992). The essential fatty acid requirement of milkfish (*Chanos chanos Forsskal*). *Fish Physiology and Biochemistry*, 9(5–6):401–407. https://doi.org/10.1007/BF02274221.
128. Takeuchi, T., Satoh, S. and Watanabe, T. (1983). Requirement of Tilapia nilotica for essential fatty acids. *Nihon Suisan Gakkai, Shi*, 49(7):1127–1134.
129. Tuncer, H., Harrell, R.M. and Chai, T.J. (1993). Beneficial effects of n-3HUFA enriched artemia as food for larval palmetto bass (*Morone saxatilis* × *M. chrysops*). *Aquaculture*, 110(3–4):341–359. https://doi.org/10.1016/0044–8486(93)90381–8.
130. Watanabe, T., Takeuchi, T., Saito, M. and Nishimura, K. (1984). Effect of low protein-high calory of essential fatty acid deficiency

diet on reproduction of rainbow trout. Nihon Suisan Gakkai, Shi; Tokyo, 50(7):1207. https://doi.org/10.2331/suisan.50.1207.

131. Castell, J.D., Sinnhuber, R.O., Wales, J.H. and Lee, D.J. (1972). Essential fatty acids in the diet of rainbow trout (*Salmo gairdneri*): growth, feed conversion and some gross deficiency symptoms. *Journal of Nutrition*, 102(1):77–85. https://doi.org/10.1093/jn/102.1.77.
132. Watanabe, T., Takashima, T. and C. (1975). Effect of dietary methyl linoleate on growth of rainbow trout. *Bulletin of the Japanese Society of Scientific Fisheries*, 40(2):181. https://doi.org/10.2331/suisan.40.181.
133. Yu, T.C. and Sinnhuber, R.O. (1972). Effect of dietary linolenic acid and docosahexaenoic acid on growth and fatty acid composition of rainbow trout (*Salmo gairdneri*). *Lipids*, 7(7):450–454.
134. Kiron, V., Fukuda, H., Takeuchi, T. and Watanabe, T. (1995). Essential fatty acid nutrition and defence mechanisms in rainbow trout *Oncorhynchus mykiss*. *Comparative Biochemistry and Physiology. Part A*, 111(3):361–367. https://doi.org/10.1016/0300–9629(95)00042–6.
135. Caballero, M.J., Obach, A., Rosenlund, G., Montero, D., Gisvold, M. and Izquierdo, M.S. (2002). Impact of different dietary lipid sources on growth, lipid digestibility, tissue fatty acid composition and histology of rainbow trout, *Oncorhynchus mykiss*. *Aquaculture*, 214(1–4):253–271. https://doi.org/10.1016/S0044–8486(01)00852–3.
136. Lochmann, R.T. and Gatlin, D.M. (1993). Essential fatty acid requirement of juvenile red drum (*Sciaenops ocellatus*). *Fish Physiology and Biochemistry*, 12(3):221–235. https://doi.org/10.1007/BF00004370.
137. Watanabe, T., Arakawa, T., Takeuchi, T. and Satoh, S. (1989). Comparison between eicosapentaenoic and docosahexaenoic acids in terms of essential fatty acid efficiency in juvenile striped jack *Pseudocaranx dentex*. Nihon Suisan Gakkai, Shi; Tokyo, 55(11):1989–1995. https://doi.org/10.2331/suisan.55.1989.
138. Bell, M.V., Henderson, R.J., Pirie, B.J.S. and Sargent, J.R. (1985). Effects of dietary polyunsaturated fatty acid deficiencies on mortality, growth and gill structure in the turbot, *Scophthalmus maximus*. *Journal of Fish Biology*, 26(2):181–191. https://doi.org/10.1111/j.1095–8649.1985.tb04255.x.

139. Watanabe, T., Thongrod, S., Takeuchi, T., Satoh, S., Kubota, S.S., Fujimaki, Y. and Cho, C.Y. (1989). Effect of dietary n-6 and n-3 fatty acids on growth, fatty acid composition and histological changes of white fish *Coregonus lavaretus maraena. Nippon Suisan Gakkaishi*, 55(11):1977–1982. https://doi.org/10.2331/suisan.55.1977.
140. Zhang, W., Wang, F., Tan, B., Yang, Q., Chi, S., Dong, X., Wang, H., Liu, H. and Zhang, S. (2019). Effects of dietary n-3HUFA on different growth stages of the white shrimp, *Litopenaeus vannamei*: growth, haematological characteristics, enzyme activities and fatty acid profiles. *Aquaculture Nutrition*, 25(5):1098–1114. https://doi.org/10.1111/anu.12926.
141. Takeuchi, T. (1997). Essential fatty acid requirements of aquatic animals with emphasis on fish larvae and fingerlings. *Reviews in Fisheries Science*, 5(1):1–25. https://doi.org/10.1080/10641269709388592.
142. Kaur, N., Chugh, V. and Gupta, A.K. (2014). Essential fatty acids as functional components of foods-a review. *Journal of Food Science and Technology*, 51(10):2289–2303. https://doi.org/10.1007/s13197–012–0677–0.
143. Tocher, D.R. (2015). Omega-3 long-chain polyunsaturated fatty acids and aquaculture in perspective. *Aquaculture*, 449:94–107. https://doi.org/10.1016/j.aquaculture.2015.01.010.
144. Jacobsen, C., Hyldig, G. and Baron, C.P. (2011). Feed composition and oxidative stability of rainbow trout during frozen storage. World. *Aquaculture*, 42(4):29–32.
145. Domínguez, R., Pateiro, M., Gagaoua, M., Barba, F.J., Zhang, W. and Lorenzo, J.M. (2019). A comprehensive review on lipid oxidation in meat and meat products. *Antioxidants*, 8(10):429; doi:10.3390/antiox8100429 https://doi.org/10.3390/antiox8100429.
146. Kubiriza, G.K., Árnason, J., Sigurgeirsson, Ó., Hamaguchi, P., Snorrason, S., Tómasson, T. and Thorarensen, H. (2017). Dietary lipid oxidation tolerance of juvenile Arctic charr (Salvelinus alpinus) and Nile tilapia (*Oreochromis niloticus*). *Aquaculture*, 467:102–108. https://doi.org/10.1016/j.aquaculture.2016.04.006.
147. Zhong, Y., Lall, S.P. and Shahidi, F. (2008). Effects of dietary oxidized oil and vitamin E on the growth, blood parameters and body composition of juvenile Atlantic cod *Gadus morhua* (Linnaeus 1758). *Aquaculture Research*, 39(15):1647–1657. https://doi.org/10.1111/j.1365–2109.2008.02038.x.

148. Zhong, Y., Lall, S.P. and Shahidi, F. (2007). Effects of oxidized dietary oil and vitamin E supplementation on lipid profile and oxidation of muscle and liver of juvenile Atlantic cod (*Gadus morhua*). *Journal of Agricultural and Food Chemistry*, 55(15):6379–6386. https://doi.org/10.1021/jf070124w.
149. Lewis-McCrea, L.M. and Lall, S.P. (2007). Effects of moderately oxidized dietary lipid and the role of vitamin E on the development of skeletal abnormalities in juvenile Atlantic halibut (*Hippoglossus hippoglossus*). *Aquaculture*, 262(1):142–155. https://doi.org/10.1016/j.aquaculture.2006.09.024.
150. Tocher, D.R., Mourente, G., Van der Eecken, A., Evjemo, J.O., Diaz, E., Wille, M., Bell, J.G. and Olsen, Y. (2003). Comparative study of antioxidant defence mechanisms in marine fish fed variable levels of oxidised oil and vitamin E. *Aquaculture International*, 11(1–2):195–216.
151. Martins, D.A., Afonso, L.O., Hosoya, S., Lewis-McCrea, L.M., Valente, L.M. and Lall, S.P. (2007). Effects of moderately oxidized dietary lipid and the role of vitamin E on the stress response in Atlantic halibut (*Hippoglossus hippoglossus* L.). *Aquaculture*, 272(1–4):573–580. https://doi.org/10.1016/j.aquaculture.2007.08.044.
152. Hamre, K., Kolås, K., Sandnes, K., Julshamn, K. and Kiessling, A. (2001). Feed intake and absorption of lipid oxidation products in Atlantic salmon (*Salmo salar*) fed diets coated with oxidised fish oil. *Fish Physiology and Biochemistry*, 25(3):209–219. https://doi.org/10.1023/A:1022257928437.
153. Roald, S.O., Armstrong, D. and Landsverk, T. (1981). Histochemical, fluorescent and electron microscopical appearance of hepatocellular ceroidosis in the Atlantic salmon *Salmo salar* L. *Journal of Fish Diseases*, 4(1):1–14. https://doi.org/10.1111/j.1365–2761.1981.tb01105.x.
154. Peng, S., Chen, L., Qin, J.G., Hou, J., Yu, N., Long, Z., Li, E. and Ye, J. (2009). Effects of dietary vitamin E supplementation on growth performance, lipid peroxidation and tissue fatty acid composition of black sea bream (*Acanthopagrus schlegeli*) fed oxidized fish oil. *Aquaculture Nutrition*, 15(3):329–337. https://doi.org/10.1111/j.1365–2095.2009.00657.x.
155. Murai, T. and Andrews, J.W. (1974). Interactions of dietary α-tocopherol, oxidized menhaden oil and ethoxyquin

on channel catfish (*Ictalurus punctatus*). *Journal of Nutrition*, 104(11):1416–1431. https://doi.org/10.1093/jn/104.11.1416.

156. Dong, G., Zhu, X., Ren, H., Nie, B., Chen, L., Li, H. and Yan, B. (2014). Effects of oxidized fish oil intake on tissue lipid metabolism and fatty acid composition of channel catfish (*Ictalurus punctatus*). *Aquaculture Research*, 45(11):1867–1880. https://doi.org/10.1111/are.12138.
157. Dong, G.F., Huang, F., Zhu, X.M., Zhang, L., Mei, M.X., Hu, Q.W. and Liu, H.Y. (2012). Nutriphysiological and cytological responses of juvenile channel catfish (*Ictalurus punctatus*) to dietary oxidized fish oil. *Aquaculture Nutrition*, 18(6):673–684. https://doi.org/10.1111/j.1365–2095.2011.00931.x.
158. Dong, X.L., Lei, W., Zhu, X.M., Han, D., Yang, Y.X. and Xie, S.Q. (2011). Effects of dietary oxidized fish oil on growth performance and skin colour of Chinese longsnout catfish (*Leiocassis longirostris* Günther). *Aquaculture Nutrition*, 17(4):e861–e868. https://doi.org/10.1111/j.1365–2095.2011.00854.x.
159. Forster, I., Higgs, D.A., Bell, G.R., Dosanjh, B.S. and March, B.E. (1988). Effect of diets containing herring oil oxidized to different degrees on growth and immunocompetence of juvenile Coho salmon (*Oncorhynchus kisutch*). *Canadian Journal of Fisheries and Aquatic Sciences*, 45(12):2187–2194. https://doi.org/10.1139/f88–254.
160. Hata, K. and Kaneda, T. (1980). Effect of autoxidized oil on carp. *Bulletin of the Japanese Society of Scientific Fisheries*, 46(8):997–1000. https://doi.org/10.2331/suisan.46.997.
161. Aoe, H., Abe, I., Saito, T., Fakawa, H. and Koyama, H. (1972). Preventive effects of tocols on muscular dystrophy of young carp. *Boll. Jap. Soc. Fish. Sci.*, 38(8):845–851. https://doi.org/10.2331/suisan.38.845.
162. Liu, W., Yang, Y., Zhang, J., Gatlin, D.M., Ringø, E. and Zhou, Z. (2014). Effects of dietary microencapsulated sodium butyrate on growth, intestinal mucosal morphology, immune response and adhesive bacteria in juvenile common carp (*Cyprinus carpio*) pre-fed with or without oxidised oil. *British Journal of Nutrition*, 112(1):15–29. https://doi.org/10.1017/S0007114514000610.
163. Ye, Y. (2016). Effects of oxidized fish oil on oxidative stress pathways of intestinal mucosa of grass carp (*Ctenopharyngodon idellus*). *Acta Hydrobiologica Sinica*/Shuisheng Shengwu Xuebao, 40(4): 758–766.

164. Chen, K. (2016). Damage of oxidized fish oil on the structure and function of hepatopancreas of grass carp (*Ctenopharyngodon idellus*). *Acta Hydrobiologica Sinica*/Shuisheng Shengwu Xuebao, 40(4):793–803.
165. Chen, K., Ye, Y. and Cai, C. (2016). Effects of dietary oxidized fish oil on the intestinal structure and permeability of grass carp (*Ctenopharyngodon idellus*). *Acta Hydrobiologica Sinica*/Shuisheng Shengwu Xuebao, 40(4):804–813.
166. Obach, A., Quentel, C. and Laurencin, F.B. (1993). Effects of alpha-tocopherol and dietary oxidized fish oil on the immune response of sea bass *Dicentrarchus labrax*. *Diseases of Aquatic Organisms*, 15:175–185. https://doi.org/10.3354/dao015175.
167. Pearson, M.D., Chinabut, S., Karnchanakharn, S. and Somsiri, T. (1994). Jaundice disease in the farmed catfish hybrid, *Clarias macrocephalus* (Gunther) × *C. gariepinus* (Burchell), in Thailand. *Journal of Fish Diseases*, 17(4):325–336. https://doi.org/10.1111/j.1365-2761.1994.tb00228.x.
168. Chen, Y.J., Liu, Y.J., Tian, L.X., Niu, J., Liang, G.Y., Yang, H.J., Yuan, Y. and Zhang, Y.Q. (2013). Effect of dietary vitamin E and selenium supplementation on growth, body composition, and antioxidant defense mechanism in juvenile largemouth bass (*Micropterus salmoides*) fed oxidized fish oil. *Fish Physiology and Biochemistry*, 39(3):593–604. https://doi.org/10.1007/s10695-012-9722-1.
169. Chen, Y.J., Liu, Y.J., Yang, H.J., Yuan, Y., Liu, F.J., Tian, L.X., Liang, G.Y. and Yuan, R.M. (2012). Effect of dietary oxidized fish oil on growth performance, body composition, antioxidant defence mechanism and liver histology of juvenile largemouth bass *Micropterus salmoides*. *Aquaculture Nutrition*, 18(3):321–331. https://doi.org/10.1111/j.1365-2095.2011.00900.x.
170. Yuan, Y., Chen, Y.J., Liu, Y.J., Yang, H.J., Liang, G.Y. and Tian, L.X. (2014). Dietary high level of vitamin premix can eliminate oxidized fish oil-induced oxidative damage and loss of reducing capacity in juvenile largemouth bass (*Micropterus salmoides*). *Aquaculture Nutrition*, 20(2):109–117. https://doi.org/10.1111/anu.12057.
171. Fontagné-Dicharry, S., Lataillade, E., Surget, A., Larroquet, L., Cluzeaud, M. and Kaushik, S. (2014). Antioxidant defense system is altered by dietary oxidized lipid in first-feeding rainbow trout (*Oncorhynchus mykiss*). *Aquaculture*, 424:220–227. https://doi.org/10.1016/j.aquaculture.2014.01.009.

172. Prearo, M.I. and Ghittino, C. (1993). Case of lipoid liver degeneration in rainbow trout cultured in Iran: evaluation of the different histological methods to detect ceroid (Caso di degenerazione lipoidea epatica in trote Irideae allevate in Iran: valutazione delle diverse metodiche istologiche per evidenziare il ceroide). *Bollettino Societa Italiana di Patologia Ittica*, 5(12):46–53.
173. Smith, C.E. (1979). The prevention of liver lipoid degeneration (ceroidosis) and microcytic anaemia in rainbow trout Salmo gairdneri Richardson fed rancid diets: a preliminary report. *Journal of Fish Diseases*, 2(5):429–437. https://doi.org/10.1111/j.1365–2761.1979.tb00394.x.
174. Cowey, C.B., Degener, E., Tacon, A.G.J., Youngson, A. and Bell, J.G. (1984). The effect of vitamin E and oxidized fish oil on the nutrition of rainbow trout (*Salmo gairdneri*) grown at natural, varying water temperatures. *British Journal of Nutrition*, 51(3):443–451. https://doi.org/10.1079/BJN19840050.
175. Moccia, R.D., Hung, S.S.O., Slinger, S.J. and Ferguson, H.W. (1984). Effect of oxidized fish oil, vitamin E and ethoxyquin on the histopathology and haematology of rainbow trout, *Salmo gairdneri* Richardson. *Journal of Fish Diseases*, 7(4):269–282. https://doi.org/10.1111/j.1365–2761.1984.tb00932.x.
176. Hung, S.S., Cho, C.Y. and Slinger, S.J. (1981). Effect of oxidized fish oil, DL-α-tocopheryl acetate and ethoxyquin supplementation on the vitamin E nutrition of rainbow trout (*Salmo gairdneri*) fed practical diets. *Journal of Nutrition*, 111(4):648–657. https://doi.org/10.1093/jn/111.4.648.
177. Daskalov, H., Robertson, P.W. and Austin, B. (2000). Influence of oxidized lipids in diets on the development of rainbow trout fry syndrome. *Journal of Fish Diseases*, 23(1):7–14. https://doi.org/10.1046/j.1365–2761.2000.00215.x.
178. Fontagné-Dicharry, S., Lataillade, E., Surget, A., Larroquet, L., Cluzeaud, M. and Kaushik, S. (2014). Antioxidant defense system is altered by dietary oxidized lipid in first-feeding rainbow trout (*Oncorhynchus mykiss*). *Aquaculture*, 424:220–227. https://doi.org/10.1016/j.aquaculture.2014.01.009.
179. Gao, J., Koshio, S., Ishikawa, M., Yokoyama, S., Nguyen, B.T. and Mamauag, R.E. (2013). Effect of dietary oxidized fish oil and vitamin C supplementation on growth performance and reduction of oxidative stress in Red Sea Bream *Pagrus major*. *Aquaculture*

Nutrition, 19(1):35–44. https://doi.org/10.1111/j.1365–2095.2011.00921.x.

180. Gao, J., Koshio, S., Ishikawa, M., Yokoyama, S., Mamauag, R.E.P. and Han, Y. (2012). Effects of dietary oxidized fish oil with vitamin E supplementation on growth performance and reduction of lipid peroxidation in tissues and blood of red sea bream *Pagrus major*. *Aquaculture*, 356:73–79. https://doi.org/10.1016/j.aquaculture.2012.05.034.
181. Fatima, M., Afzal, M. and Shah, S.Z.H. (2019). Effect of dietary oxidized oil and vitamin E on growth performance, lipid peroxidation and fatty acid profile of *Labeo rohita* fingerlings. *Aquaculture Nutrition*, 25(2):281–291. https://doi.org/10.1111/anu.12851.
182. Boglino, A., Darias, M.J., Estévez, A., Andree, K.B., Sarasquete, C., Ortiz-Delgado, J.B., Solé, M. and Gisbert, E. (2014). The effect of dietary oxidized lipid levels on growth performance, antioxidant enzyme activities, intestinal lipid deposition and skeletogenesis in Senegalese sole (*Solea senegalensis*) larvae. *Aquaculture Nutrition*, 20(6):692–711. https://doi.org/10.1111/anu.12123.
183. Fontagné, S., Bazin, D., Brèque, J., Vachot, C., Bernarde, C., Rouault, T. and Bergot, P. (2006). Effects of dietary oxidized lipid and vitamin A on the early development and antioxidant status of Siberian sturgeon (*Acipenser baeri*) larvae. *Aquaculture*, 257(1–4):400–411. https://doi.org/10.1016/j.aquaculture.2006.01.025.
184. Pearse, L., MacQueen, A. and Roberts, R.J. (1974). Muscular dystrophy in cultured turbot (*Scophthalmus maximus* L). *Veterinary Record*, 94(19):435–437. https://doi.org/10.1136/vr.94.19.435.
185. Obach, A. and Laurencin, F.B. (1992). Effects of dietary oxidized fish oil and deficiency of anti-oxidants on the immune response of turbot, *Scophthalmus maximus*. *Aquaculture*, 107(2–3):221–228. https://doi.org/10.1016/0044–8486(92)90070–2.
186. Yang, S.P., Liu, H.L., Wang, C.G., Yang, P., Sun, C.B. and Chan, S.M. (2015). Effect of oxidized fish oil on growth performance and oxidative stress of *Litopenaeus vannamei*. *Aquaculture Nutrition*, 21(1):121–127. https://doi.org/10.1111/anu.12143.
187. Murai, T., Akiyama, T., Ogata, H. and Suzuki, T. (1988). Interaction of dietary oxidized fish oil and glutathione on fingerling yellowtail *Seriola quinqueradiata*. *Nippon Suisan Gakkaishi*, *54*(1):145–149. https://doi.org/10.2331/suisan.54.145.

188. Zhuo, L. (2018). *Influence of oxidized fish oil on the intestinal health of juvenile yellow catfish (Pelteobagrus fulvidraco) and the use of arginine as an intervention measure.* Shui Chan Xue Bao, Shanghai, 42(1), pp. 100–111.
189. Soliman,A.K., Roberts, R.J. and Jauncey, K. (1983). The pathological effects of feeding rancid lipid in diets for *Oreochromis niloticus* (Trewavas). In *Proceedings of the International Symposium in Tilapia in Aquaculture,* Nazareth, Israel, May 8–13 1983. L. Fishelson et al. (Eds.), pp. 193–199. Tel Aviv University Press, Tel Aviv.
190. Lundebye, A.K., Hove, H., Måge, A., Bohne, V.J.B. and Hamre, K. (2010). Levels of synthetic antioxidants (ethoxyquin, butylated hydroxytoluene and butylated hydroxyanisole) in fish feed and commercially farmed fish. *Food Additives and Contaminants: Part A,* 27(12): 1652–1657. https://doi.org/10.1080/19440049.2010.508195.
191. Augustyniak, A., Bartosz, G., Čipak, A., Duburs, G., Horáková, L.U. et al. (2010). Natural and synthetic antioxidants: an updated overview. *Free Radical Research,* 44(10):1216–1262. https://doi.org/10.3109/10715762.2010.508495.
192. Ørnsrud, R., Arukwe, A., Bohne, V., Pavlikova, N. and Lundebye, A.K. (2011). Investigations on the metabolism and potentially adverse effects of ethoxyquin dimer, a major metabolite of the synthetic antioxidant ethoxyquin in salmon muscle. *Journal of Food Protection,* 74(9):1574–1580.
193. Kubiriza, G.K., Árnarson, J., Sigurgeirsson, Ó., Hamaguchi, P., Snorrason, S., Tómasson, T. and Thorarensen, H. (2019). Growth and hepatic antioxidant enzyme activity of juvenile Arctic charr (*Salvelinus alpinus*) fed on diets supplemented with ethoxyquin, rosemary (*Rosmarinus officinalis*), or bladder wrack (*Fucus vesiculosus*). *Aquaculture International,* 27(1):287–301. https://doi.org/10.1007/s10499-018-0324-5.
194. Li, H., Tang, S., Du, W., Jiang, J., Peng, P., Yuan, P., Liao, Y., Long, J. and Zhou, S. (2019). The effects of ethoxyquin and Angelica sinensis extracts on lipid oxidation in fish feeds and growth, digestive and absorptive capacities and antioxidant status in juvenile red carp (*Cyprinus carpio* var. *xingguonensis*): a comparative study. *Fish Physiology and Biochemistry,* 45(1):43–61. https://doi.org/10.1007/s10695-018-0533-x.
195. Gul, K., Yousuf, B., Singh, A.K., Singh, P. and Wani, A.A. (2015). Rice bran: nutritional values and its emerging potential for

development of functional food-A review. *Bioactive Carbohydrates and Dietary Fibre*, 6(1):24–30. https://doi.org/10.1016/j.bcdf.2015.06.002.

196. Wasowicz, E., Gramza, A., Hêœ, M., Jeleń, H.H. and Korczak, J. Ma³ecka, M. Mildner-Szkudlarz, S., Rudzińska, M., Samotyja, U. and Zawirska-Wojtasiak, R (2004). Oxidation of lipids in food. *Polish Journal of Food and Nutrition Sciences*, 13(54):87–100.
197. Davis, D.A. and Gatlin 3rd, D.M. (1996). Dietary mineral requirements of fish and marine crustaceans. *Reviews in Fisheries Science*, 4(1):75–99. https://doi.org/10.1080/10641269609388579.
198. Antony Jesu Prabhu, P., Schrama, J.W. and Kaushik, S.J. (2016). Mineral requirements of fish: a systematic review. *Reviews in Aquaculture*, 8(2):172–219. https://doi.org/10.1111/raq.12090.
199. Chanda, S., Paul, B.N., Ghosh, K. and Giri, S.S. (2015). Dietary essentiality of trace minerals in aquaculture-A Review. *Agricultural Reviews*, 36(2):100–112. https://doi.org/10.5958/0976–0741.2015.00012.4.
200. Watanabe, T., Kiron, V. and Satoh, S. (1997). Trace minerals in fish nutrition. *Aquaculture*, 151(1–4):185–207. https://doi.org/10.1016/S0044–8486(96)01503–7.
201. Poulsen, H.D. (2005). Trace elements in animal nutrition. In *Essential trace elements for plants, animals and humans.* NJF Seminar No. 370, Rekjavik, Iceland, August 15th–17th 2005. 3, pp. 34–36, Agricultural Universityof Iceland, Rit LBHI nr.
202. Vielma, J. and Lall, S.P. (1998). Phosphorus utilization by Atlantic salmon (*Salmo salar*) reared in freshwater is not influenced by higher dietary calcium intake. *Aquaculture*, 160(1–2):117–128. https://doi.org/10.1016/S0044–8486(97)00300–1.
203. O'Connell, J.P. and Gatlin 3rd, D.M. (1994). Effects of dietary calcium and vitamin D3 on weight gain and mineral composition of the blue tilapia (*Oreochromis aureus*) in low-calcium water. *Aquaculture*, 125(1–2):107–117. https://doi.org/10.1016/0044–8486(94)90287–9.
204. Robinson, E.H., LaBomascus, D., Brown, P.B. and Linton, T.L. (1987). Dietary calcium and phosphorus requirements of *Oreochromis aureus* reared in calcium-free water. *Aquaculture*, 64(4):267–276. https://doi.org/10.1016/0044–8486(87)90189-X.
205. Robinson, E.H., Rawles, S.D., Yette, H.E. and Greene, L.W. (1984). An estimate of the dietary calcium requirement of fingerling

Tilapia aurea reared in calcium-free water. *Aquaculture*, 41(4): 389–393. https://doi.org/10.1016/0044–8486(84)90206–0.

206. Robinson, E.H., Rawles, S.D., Brown, P.B., Yette, H.E. and Greene, L.W. (1986). Dietary calcium requirement of channel catfish *Ictalurus punctatus*, reared in calcium-free water. *Aquaculture*, 53(3–4):263–270. https://doi.org/10.1016/0044–8486(86)90356-X.
207. Gatlin 3rd, D.M. and Phillips, H.F. (1989). Dietary calcium, phytate and zinc interactions in channel catfish. *Aquaculture*, 79(1–4): 259–266. https://doi.org/10.1016/0044–8486(89)90466–3.
208. Liang, J.-J., Liu, Y.-J., Yang, Z.-N., Tian, L.-X., Yang, H.-J. and Liang, G.-Y. (2912). Dietary calcium requirement and effects on growth and tissue calcium content of juvenile grass carp (*Ctenopharyngodon idella*). *Aquaculture Nutrition*, 18(5):544–550. https://doi.org/10.1111/j.1365–2095.2011.00916.x.
209. Shiau, S.-Y. and Tseng, H.-C. (2007). Dietary calcium requirements of juvenile tilapia, *Oreochromis niloticus* × *O. aureus*, reared in fresh water. *Aquaculture Nutrition*, 13(4):298–303. https://doi.org/10.1111/j.1365–2095.2007.00481.x.
210. Hossain, M.A. and Furuichi, M. (2000). Necessity of calcium supplement to the diet of Japanese flounder. *Fisheries Science*, 66(4): 660–664. https://doi.org/10.1046/j.1444–2906.2000.00130.x.
211. Lu, J., Yoshizaki, G., Endo, M. and Takeuchi, T. (2013). Effect of dietary high amount of calcium and phosphorous on reducing the prevalence of morphological deformities in GH-transgenic Nile tilapia. *Fisheries Science*, 79(4):647–658. https://doi.org/10.1007/s12562–013–0626–5.
212. Ye, C.X., Liu, Y.J., Tian, L.X., Mai, K.S., Du, Z.Y., Yang, H.J. and Niu, J. (2006). Effect of dietary calcium and phosphorus on growth, feed efficiency, mineral content and body composition of juvenile grouper, *Epinephelus coioides*. *Aquaculture*, 255(1–4):263–271. https://doi.org/10.1016/j.aquaculture.2005.12.028.
213. Fontagné, S., Silva, N., Bazin, D., Ramos, A., Aguirre, P., Surget, A., Abrantes, A., Kaushik, S.J. and Power, D.M. (2009). Effects of dietary phosphorus and calcium level on growth and skeletal development in rainbow trout (*Oncorhynchus mykiss*) fry. *Aquaculture*, 297(1–4):141–150. https://doi.org/10.1016/j.aquaculture.2009.09.022.
214. Hossain, M.A.R. and Furuichi, M. (2000). Essentiality of dietary calcium supplement in redlip mullet, *Liza haematocheila*. *Aquaculture*

Nutrition, 6(1):33–38. https://doi.org/10.1046/j. 1365–2095.2000. 00123.x.

215. Hossain, M.A. and Furuichi, M. (2000). Effects of calcium supplement to purified diets on growth and bone mineralization in juvenile red sea bream and black sea bream. *Journal of the Faculty of Agriculture Kyushu University*, 45(2):481–486. https://doi.org/10.5109/24398.
216. Hossain, M.A. and Furuichi, M. (1998). Availability of environmental and dietary calcium in tiger puffer. *Aquaculture International*, 6(2):121–132. https://doi.org/10.1023/A:1009286206931.
217. Huang, K., Wang, W., Kong, L. and Huang, H. (2004). Requirements of *Penaeus vannamei* in low salinity water for dietary phosphorus and calcium. *Journal of Ocean University of Qingdao*, 34(2):209–216.
218. Ketola, H.G. Requirement of Atlantic salmon for dietary phosphorus. *Transactions of the American Fisheries Society*, 104(3):548–551. https://doi.org/10.1577/1548–8659(1975)104<548:ROASFD>2.0.CO;2.
219. Åsgård, T. and Shearer, K.D. (1997). Dietary phosphorus requirement of juvenile Atlantic salmon, *Salmo salar* L. *Aquaculture Nutrition*, 3(1):17–23. https://doi.org/10.1046/j.1365–2095.1997.00069.x.
220. Shao, Q., Ma, J., Xu, Z., Hu, W., Xu, J. and Xie, S. (2008). Dietary phosphorus requirement of juvenile black seabream, *Sparus macrocephalus*. *Aquaculture*, 277(1–2):92–100. https://doi.org/10.1016/j.aquaculture.2008.01.029.
221. Sukumaran, K., Pal, A.K., Sahu, N.P., Debnath, D. and Patro, B. (2009). Phosphorus requirement of catla (*catla catla* Hamilton) fingerlings based on growth, whole-body phosphorus concentration and non-faecal phosphorus excretion. *Aquaculture Research*, 40(2): 139–147. https://doi.org/10.1111/j.1365–2109.2008.02073.x.
222. Nwanna, L.C., Kühlwein, H. and Schwarz, F.J. (2010). Phosphorus requirement of common carp (*Cyprinus carpio* L) based on growth and mineralization. *Aquaculture Research*, 41(3):401–410. https://doi.org/10.1111/j.1365–2109.2009.02221.x.
223. Lovell, R.T. (1978). Dietary phosphorus requirement of Channel catfish (*Ictalurus punctatus*). *Transactions of the American Fisheries Society*, 107(4):617–621.
224. Yuan, Y.C., Yang, H.J., Gong, S.Y., Luo, Z., Yu, D.H., Yan, J.L. and Yang, X.F. (2011). Dietary phosphorus requirement of juvenile Chinese sucker, *Myxocyprinus asiaticus*. *Aquaculture Nutrition*, 17(2):159–169.

225. Watanabe, T., Murakami, A., Takeuch, L., Nose, T. and Ogino, C. (1980). Requirement of chum salmon held in freshwater for dietary phosphorus. Nihon Suisan Gakkai, Shi; Tokyo, 46(3):361–367. https://doi.org/10.2331/suisan.46.361.
226. Sun, Y., Li, B., Zhang, X., Chen, M., Tang, H. and Yu, X. (2018). Dietary available phosphorus requirement of crucian carp, *Carassius auratus. Aquaculture Nutrition*, 24(5):1494–1501.
227. Oliva-Teles, A. and Pimentel-Rodrigues, A. (2004). Phosphorus requirement of European sea bass (*Dicentrarchus labrax* L.) juveniles. *Aquaculture Research*, 35(7):636–642. https://doi.org/10.1111/j.1365–2109.2004.01059.x.
228. Xie, D., Zhu, X., Yang, Y., Han, D., Jin, J. and Xie, S. (2018). Dietary available phosphorus requirement for juvenile gibel carp (*Carassius auratus gibelio* var. *CASIII*). *Aquaculture Research*, 49(3):1284–1292. https://doi.org/10.1111/are.13582.
229. Xie, D., Han, D., Zhu, X., Yang, Y., Jin, J., Liu, H. and Xie, S. (2017). Dietary available phosphorus requirement for on-growing gibel carp (*Carassius auratus gibelio* var. *CASIII*). *Aquaculture Nutrition*, 23(5):1104–1112. https://doi.org/10.1111/anu.12478.
230. Pimentel-Rodrigues, A.M. and Oliva-Teles, A. (2001). Phosphorus requirements of gilthead sea bream (*Sparus aurata* L.) juveniles. *Aquaculture Research*, 32:157–161. https://doi.org/10.1046/j.1355–557x.2001.00013.x.
231. Liang, J.-J., Liu, Y.-J., Tian, L.-X., Yang, H.-J. and Liang, G.-Y. (2012). Dietary available phosphorus requirement of juvenile grass carp (*Ctenopharyngodon idella*). *Aquaculture Nutrition*, 18(2):181–188. https://doi.org/10.1111/j.1365–2095.2011.00887.x.
232. Musharraf, M. and Khan, M.A. (2019). Dietary phosphorus requirement of fingerling Indian major carp, *Labeo rohita* (Hamilton). *Journal of the World Aquaculture Society*, 50(2):469–484. https://doi.org/10.1111/jwas.12521.
233. Zhang, C., Mai, K., Ai, Q., Zhang, W., Duan, Q., Tan, B., Ma, H., Xu, W., Liufu, Z. and Wang, X. (2006). Dietary phosphorus requirement of juvenile Japanese seabass, *Lateolabrax japonicus. Aquaculture*, 255(1–4):201–209. https://doi.org/10.1016/j.aquaculture.2005.11.040.
234. Roy, P.K. and Lall, S.P. (2003). Dietary phosphorus requirement of juvenile haddock (*Melanogrammus aeglefinus* L.). *Aquaculture*,

221(1–4):451–468. https://doi.org/10.1016/S0044–8486(03)0006 5–6.

235. Borlongan, I.G. and Satoh, S. (2001). Dietary phosphorus requirement of juvenile milkfish, *Chanos chanos* (Forsskal). *Aquaculture Research,* 32:26–32. https://doi.org/10.1046/j.1355–557x.2001. 00003.x.
236. Kandeepan, C. and Arunachalam, S. (2005). Dietary phosphorus requirement of *Oreochromis mossambicus. Journal of Ecobiology,* 17(6):571–579.
237. Paul, B.N., Sarkar, S., Giri, S.S., Rangacharyulu, P.V. and Mohanty, S.N. (2004). Phosphorus requirements and optimum calcium/phosphorus ratio in the diet of *mrigal Cirrhinus mrigala* (Ham.) fingerlings. *Journal of Applied Ichthyology,* 20(4):306–309. https://doi.org/10.1111/j.1439–0426.2004.00541.x.
238. Yao, Y.F., Jiang, M., Wen, H., Wu, F., Liu, W., Tian, J. and Yang, C.G. (2014). Dietary phosphorus requirement of GIFT strain of Nile tilapia *Oreochromis niloticus* reared in freshwater. *Aquaculture Nutrition,* 20(3):273–280. https://doi.org/10.1111/anu.12075.
239. Choi, S.M., Kim, K.W., Kang, Y.J., Wang, X., Kim, J.W., Yoo, G.Y. and Bai, S.C. (2005). Reevaluation of the phosphorus requirement of juvenile olive flounder *Paralichfhys olivaceus* and the bioavailability of various inorganic phosphorus sources. *Journal of the World Aquaculture Society,* 36(2):217–222. https://doi.org/10.1111/j.1749–7345.2005.tb00388.x.
240. Rodehutscord, M. and Pfeffer, E. (1995). Requirement for phosphorus in rainbow trout (*Oncorhynchus mykiss*) growing from 50 to 200 g. *Water Science and Technology,* 31(10):137–141. https://doi.org/10.2166/wst.1995.0370.
241. Ketola, H.G. and Richmond, M.E. (1994). Requirement of rainbow trout for dietary phosphorus and its relationship to the amount discharged in hatchery effluents. *Transactions of the American Fisheries Society,* 123(4):587–594. https://doi.org/10.1577/1548–8659(1994)123<0587:RORTFD>2.3.CO;2.
242. Ogino, C. and Takeda, H. (1978). Requirements of rainbow trout for dietary calcium and phosphorus. Nihon Suisan Gakkai, Shi; Tokyo, 44(9):1019. https://doi.org/10.2331/suisan.44.1015.
243. El Zibdeh, M., Yoshimatsu, T., Matsui, S., Furuichi, M., Kitajima, C. and Azuma, R. (1995). Requirement of redlip mullet [*Liza hematochiela*] for dietary phosphorus. *Journal of the Faculty of*

Agriculture Kyushu University, 40(1–2):135–145. https://doi.org/10.5109/24100.

244. Xu, Q.Y., Xu, H., Wang, C., Zheng, Q. and Sun, D. (2011). Studies on dietary phosphorus requirement of juvenile Siberian sturgeon *Acipenser baerii*. *Journal of Applied Ichthyology*, 27(2):709–714. https://doi.org/10.1111/j.1439–0426.2011.01677.x.
245. Dougall, D.S., Woods 3rd, L.C., Douglass, L.W. and Soares, J.H. (1996). Dietary phosphorus requirement of juvenile striped bass *Morone saxatilis* 1. *Journal of the World Aquaculture Society*, 27(1): 82–91. https://doi.org/10.1111/j.1749–7345.1996.tb00597.x.
246. Brown, M.L., Jaramillo Jr, F. and Gatlin 3rd, D.M. (1993). Dietary phosphorus requirement of juvenile sunshine bass, *Morone chrysops*♀ × *M. saxatilis*♂. *Aquaculture*, 113(4):355–363. https://doi.org/10.1016/0044–8486(93)90405-N.
247. Zafar, N. and Khan, M.A. (2018). Determination of dietary phosphorus requirement of stinging catfish *Heteropneustes fossilis* based on feed conversion, growth, vertebrae phosphorus, whole body phosphorus, haematology and antioxidant status. *Aquaculture Nutrition*, 24(5):1577–1586. https://doi.org/10.1111/anu.12794.
248. Araújo, J.G., Guimarães, I.G., Mota, C.S., De Paula, F.G., Café, M.B. and Pádua, D.M.C. (2017). Dietary available phosphorus requirement for tambaqui, *Colossoma macropomum*, juveniles based on growth, haematology and bone mineralization. *Aquaculture Nutrition*, 23(4):822–832. https://doi.org/10.1111/anu.12449.
249. Yu, H.R., Zhang, Q., Xiong, D.M., Huang, G.Q., Li, W.Z. and Liu, S.W. (2013). Dietary available phosphorus requirement of juvenile walking catfish, *Clarias leather*. *Aquaculture Nutrition*, 19(4): 483–490. https://doi.org/10.1111/j.1365–2095.2012.00982.x.
250. Luo, Z., Tan, X.Y., Liu, X. and Wang, W.M. (2010). Dietary total phosphorus requirement of juvenile yellow catfish *Pelteobagrus fulvidraco*. *Aquaculture International*, 18(5):897–908. https://doi.org/10.1007/s10499–009–9310–2.
251. Mai, K., Zhang, C., Ai, Q., Duan, Q., Xu, W., Zhang, L., Liufu, Z. and Tan, B. (2006). Dietary phosphorus requirement of large yellow croaker, *Pseudosciaena crocea* R. *Aquaculture*, 251(2–4):346–353. https://doi.org/10.1016/j.aquaculture.2005.05.038.
252. Huang, K., Wang, W., Kong, L. and Huang, H. (2004). Requirements of Penaeus vannamei in low salinity water for dietary phosphorus and calcium. Journal of Ocean University of Qingdao,

2. *Periodical of Ocean University of China*/Zhongguo Haiyang Daxue Xuebao, 34(2):209–216.

253. El-Zibdeh, M., Ide, K., Yoshimatsu, T., Matsui, S. and Furuichi, M. (1995). Requirement of yellow croaker *Nibea albiflora* for dietary phosphorus. *Journal of the Faculty of Agriculture Kyushu University*, 40(1/2):147–156. https://doi.org/10.5109/24101.
254. Shearer, K.D. (1988). Dietary potassium requirement of juvenile chinook salmon. *Aquaculture*, 73(1–4):119–129. https://doi.org/10.1016/0044–8486(88)90047–6.
255. Chen, C., Zhu, W., Wu, F., Liu, M., Tan, Q., Han, D., Xie, S. and Liang, X. (2016). Quantifying the dietary potassium requirement of subadult grass carp (*Ctenopharyngodon idellus*). *Aquaculture Nutrition*, 22(3):541–549. https://doi.org/10.1111/anu.12279.
256. Zhu, W., Liu, M., Chen, C., Wu, F., Yang, J., Tan, Q., Xie, S. and Liang, X. (2014). Quantifying the dietary potassium requirement of juvenile grass carp (*Ctenopharyngodon idellus*). *Aquaculture*, 430:218–223. https://doi.org/10.1016/j.aquaculture.2014.04.014.
257. Liang, J.J., Yang, H.J., Liu, Y.J. and Tian, L.X. (2014). Dietary potassium requirement of juvenile grass carp (*Ctenopharyngodon idella* Val.) based on growth and tissue potassium content. *Aquaculture Research*, 45(4):701–708. https://doi.org/10.1111/are.12008.
258. Musharraf, M. and Khan, M.A. (2019). Dietary potassium requirement of fingerling Indian major carp, *Labeo rohita* (Hamilton), based on growth parameters, gill Na+–K+ ATPase activity, tissue mineralization and antioxidant activities. *Aquaculture Nutrition*, 25(2):271–280. https://doi.org/10.1111/anu.12850.
259. Shiau, S.Y. and Hsieh, J.F. (2001). Dietary potassium requirement of juvenile grass shrimp *Penaeus monodon*. *Fisheries Science*, 67(4): 592–595. https://doi.org/10.1046/j.1444–2906.2001.00294.x.
260. Fa-yi, L., He-fang, L., Hui-Liang, W., De-hai, L. and Yu-chuan, T. (1995). Requirements of shrimp. *Chinese Journal of Oceanology and Limnology*, 13(2):141–146. https://doi.org/10.1007/BF02846819.
261. Kanazawa, A., Teshima, S. and Sasaki, M. (1984). Requirements of the juvenile prawn for calcium, phosphorus, magnesium, potassium, copper, manganese, and iron. *Memoirs of Faculty of Fisheries Kagoshima University*, 33(1):63–71.
262. El-Mowafi, A.F.A. and Maage, A. (1998). Magnesium requirement of Atlantic salmon (*Salmo salar* L.) parr in seawater-treated fresh water. *Aquaculture Nutrition*, 4(1):31–38.

263. Ogino, D. and Chiou, J.Y. (1976). Mineral requirements in fish –2. Magnesium requirement of carp. *Bulletin of the Japanese Society of Scientific Fisheries*, 42(1):71–75. https://doi.org/10.2331/suisan.42.71.
264. Liang, J.J., Tian, L.X., Liu, Y.J., Yang, H.J. and Liang, G.Y. (2012). Dietary magnesium requirement and effects on growth and tissue magnesium content of juvenile grass carp (*Ctenopharyngodon idella*). *Aquaculture Nutrition*, 18(1):56–64. https://doi.org/10.1111/j.1365–2095.2011.00876.x.
265. Wang, F.B., Luo, L., Lin, S.M., Li, Y., Chen, S., Wang, Y.G., Wen, H. and Hu, C.J. (2011). Dietary magnesium requirements of juvenile grass carp, *Ctenopharyngodon idella. Aquaculture Nutrition*, 17(3):e691–e700.
266. Lin, Y.H., Ku, C.Y. and Shiau, S.Y. (2013). Estimation of dietary magnesium requirements of juvenile tilapia, *Oreochromis niloticus* × *Oreochromis aureus*, reared in freshwater and seawater. *Aquaculture*, 380:47–51. https://doi.org/10.1016/j.aquaculture.2012.11.034.
267. Knox, D., Cowey, C.B. and Adron, J.W. (1981). Studies on the nutrition of salmonid fish. The magnesium requirement of rainbow trout (*Salmo gairdneri*). *British Journal of Nutrition*, 45(1):137–148. https://doi.org/10.1079/BJN19810086.
268. Ogino, C., Takashima, F. and Chiou, J.Y. (1978). Requirement of rainbow trout for dietary magnesium. Nihon Suisan Gakkai, Shi, Tokyo. 44(10):1105–1108. https://doi.org/10.2331/suisan.44.1105.
269. Cheng, K.M., Hu, C.Q., Liu, Y.N., Zheng, S.X. and Qi, X.J. (2005). Dietary magnesium requirement and physiological responses of marine shrimp *Litopenaeus vannamei* reared in low salinity water. *Aquaculture Nutrition*, 11(5):385–393. https://doi.org/10.1111/j.1365–2095.2005.00364.x.
270. Andersen, F., Maage, A. and Julshamn, K. (1996). An estimation of dietary iron requirement of Atlantic salmon, *Salmo salar* L., parr. *Aquaculture Nutrition*, 2(1):41–47. https://doi.org/10.1111/j.1365–2095.1996.tb00006.x.
271. Gatlin 3rd, D.M. and Wilson, R.P. (1986). Characterization of iron deficiency and the dietary iron requirement of fingerling channel catfish. *Aquaculture*, 52(3):191–198. https://doi.org/10.1016/0044–8486(86)90143–2.

272. Sakamoto, S. and Yone, Y. (1978). Requirement of Red Sea bream for dietary iron-II. Nihon Suisan Gakkai, Shi, Tokyo. 44(3):223 https://doi.org/10.2331/suisan.44.223.
273. Gatlin 3rd, D.M. and Wilson, R.P. (1983). Dietary zinc requirement of fingerling channel catfish. *Journal of Nutrition*, 113(3): 630–635. https://doi.org/10.1093/jn/113.3.630.
274. Paripatananont, T. and Lovell, R.T. (1995). Chelated zinc reduces the dietary zinc requirement of channel catfish, *Ictalurus punctatus*. *Aquaculture*, 133(1):73–82. https://doi.org/10.1016/0044–8486(94)00404-C.
275. Scarpa, J. and Gatlin 3rd, D.M. (1992). Dietary zinc requirements of channel catfish, *Ictalurus punctatus*, swim-up fry in soft and hard water. *Aquaculture*, 106(3–4):311–322. https://doi.org/10.1016/0044–8486(92)90263-K.
276. Ogino, C.G.-Y. *Yang*. (1979). Requirement of carp for dietary zinc. Nihon Suisan Gakkai, Shi, p. 45(8):967.
277. Liang, J.-J., Yang, H.-J., Liu, Y.-J., Tian, L.-X. and Liang, G.-Y. (2012). Dietary zinc requirement of juvenile grass carp (*Ctenopharyngodon idella*) based on growth and mineralization. *Aquaculture Nutrition*, 18(4):380–387. https://doi.org/10.1111/j.1365–2095.2011.00935.x.
278. Shiau, S.Y. and Jiang, L.C. (2006). Dietary zinc requirements of grass shrimp, *Penaeus monodon*, and effects on immune responses. *Aquaculture*, 254(1–4):476–482. https://doi.org/10.1016/j.aquaculture.2005.10.033.
279. Houng-Yung, C., Yu-Chun, C., Li-Chi, H. and Meng-Hsien, C. (2014). Dietary zinc requirements of juvenile grouper, *Epinephelus malabaricus*. *Aquaculture*, 432:360–364. https://doi.org/10.1016/j.aquaculture.2014.05.020.
280. Eid, A.E. and Ghonim, S.I. (1994). Dietary zinc requirement of fingerling *Oreochromis niloticus*. *Aquaculture*, 119(2–3):259–264. https://doi.org/10.1016/0044–8486(94)90180–5.
281. Huang, F., Jiang, M., Wen, H., Wu, F., Liu, W., Tian, J. and Yang, C. (2015). Dietary zinc requirement of adult Nile tilapia (*Oreochromis niloticus*) fed semi-purified diets, and effects on tissue mineral composition and antioxidant responses. *Aquaculture*, 439:53–59. https://doi.org/10.1016/j.aquaculture.2015.01.018.
282. Huang, Q.C., Wang, E.L., Dong, X.H., Tan, B.P., Chi, S.Y., Yang, Q.H., Zhang, S., Liu, H.Y. and Yang, Y.Z. (2018). Investigations

on zinc bioavailability of different sources and dietary zinc requirement in juvenile grouper *Epinephelus coioides*. *Aquaculture Research*, 49(8):2763–2773. https://doi.org/10.1111/are.13737.
283. Ogino, C. and Yang, G.-Y. (1978). Requirement of rainbow trout for dietary zinc. Nihon Suisan Gakkai, Shi, Tokyo., 44(9): 1015–1018. https://doi.org/10.2331/suisan.44.1015.
284. Gatlin 3rd, D.M., O'Connell, J.P. and Scarpa, J. (1991). Dietary zinc requirement of the red drum, *Sciaenops ocellatus*. *Aquaculture*, 92:259–265. https://doi.org/10.1016/0044–8486(91) 90027–5.
285. Moazenzadeh, K., Islami, H.R., Zamini, A. and Soltani, M. (2017). Dietary zinc requirement of Siberian sturgeon (*Acipenser baerii*, Brandt 1869) juveniles, based on the growth performance and blood parameters. *International Aquatic Research*, 9(1):25–35. https://doi.org/10.1007/s40071–017–0153–6.
286. Luo, Z., Tan, X.Y., Zheng, J.L., Chen, Q.L. and Liu, C.X. (2011). Quantitative dietary zinc requirement of juvenile yellow catfish *Pelteobagrus fulvidraco*, and effects on hepatic intermediary metabolism and antioxidant responses. *Aquaculture*, 319(1–2):150–155. https://doi.org/10.1016/j.aquaculture.2011.06.047.
287. Gatlin 3rd, D.M. and Wilson, R.P. (1986). Dietary copper requirement of fingerling channel catfish. *Aquaculture*, 54(4):277–285. https://doi.org/10.1016/0044–8486(86)90272–3.
288. Qiao, Y.G., Tan, B.P., Mai, K.S. and Al, Q.H. (2013). Determination of the dietary copper requirement for cobia (*Rachycentron canadum*). *Periodical of Ocean University of China*, 4.
289. Ogino, C. and Yang, G.-Y. (1980). Requirements of carp and rainbow trout for dietary manganese and copper. *Nihon Suisan Gakkai, Shi*, Tokyo, 46(4):455–458.
290. Zafar, N. and Khan, M.A. (2020). Dietary copper requirement of fingerling *Heteropneustes fossilis* for formulating copper-balanced commercial feeds. *Aquaculture Nutrition*, 26(2):248–260. https://doi.org/10.1111/anu.12986.
291. Lee, M.H. and Shiau, S.Y. (2002). Dietary copper requirement of juvenile grass shrimp, *Penaeus monodon*, and effects on non-specific immune responses. *Fish and Shellfish Immunology*, 13(4):259–270. https://doi.org/10.1006/fsim.2001.0401.
292. Lin, Y.H., Shih, C.C., Kent, M. and Shiau, S.Y. (2010). Dietary copper requirement reevaluation for juvenile grouper,

Epinephelus malabaricus, with an organic copper source. *Aquaculture*, 310(1–2):173–177. https://doi.org/10.1016/j.aquaculture.2010.10.004.

293. Ma, H.Y., Yang, Q.H., Dong, X.H., Tan, B.P., Chi, S.Y. and Liu, H.Y. (2014). Dietary requirement of juvenile grouper, *Epinephelus coioides* for hydroxy methionine copper. *Journal of Guangdong Ocean University*, 34(4):1–3.
294. Wang, H., Li, E., Zhu, H., Du, Z., Qin, J. and Chen, L. (2016). Dietary copper requirement of juvenile Russian sturgeon *Acipenser gueldenstaedtii*. *Aquaculture*, 454:118–124.
295. Abdel-Hameid, N.A.H., Zehra, S. and Khan, M.A. (2017). Dietary copper requirement of fingerling *Channa punctatus* (Bloch) based on growth, feed conversion, blood parameters and whole body copper concentration. *Aquaculture Research*, 48(6):2787–2797. https://doi.org/10.1111/are.13112.
296. Ali, S.A. (2000). Copper, manganese and zinc requirements in the diet of shrimp *Penaeus indicus*. *Asian Fisheries Science*, 13(3): 201–207.
297. Davis, D.A., Lawrence, A.L. and Gatlin 3rd, D. (1993). Dietary copper requirement of *Penaeus vannamei*. *Nippon Suisan Gakkaishi*, 59(1):117–122. https://doi.org/10.2331/suisan.59.117.
298. Tan, X.Y., Luo, Z., Liu, X. and Xie, C.X. (2011). Dietary copper requirement of juvenile yellow catfish *Pelteobagrus fulvidraco*. *Aquaculture Nutrition*, 17(2):170–176. https://doi.org/10.1111/j.1365–2095.2009.00720.x.
299. Cao, J., Miao, X., Xu, W., Li, J., Zhang, W. and Mai, K. (2014). Dietary copper requirements of juvenile large yellow croaker *Larimichthys croceus*. *Aquaculture*, 432:346–350. https://doi.org/10.1016/j.aquaculture.2014.05.032.
300. Maage, A., Lygren, B. and El-Mowafi, A.F.A. (2000). Manganese requirement of Atlantic salmon (*Salmo salar*) fry. *Fisheries Science*, 66(1):1–8. https://doi.org/10.1046/j.1444–2906.2000.00001.x.
301. Liu, K., Ai, Q.H., Mai, K.S., Zhang, W.B., Zhang, L. and Zheng, S.X. (2013). Dietary manganese requirement for juvenile cobia, *Rachycentron canadum* L. *Aquaculture Nutrition*, 19(4):461–467. https://doi.org/10.1111/j.1365–2095.2012.00979.x.
302. Liang, J.J., Wang, S., Han, B., Tian, L.X., Yang, H.J. and Liu, Y.J. (2015). Dietary manganese requirement of juvenile grass carp (*Ctenopharyngodon idella* Val.) based on growth and tissue manganese

concentration. *Aquaculture Research*, 46(12):2991–2998. https://doi.org/10.1111/are.12455.

303. Liu, Y., Wang, J.Y., Li, B.S., Qiao, H.J., Liu, X.D., Hao, T.T. and Wang, X.Y. (2018). Dietary manganese requirement of juvenile hybrid grouper, *Epinephelus lanceolatus* × *E. fuscoguttatus*. *Aquaculture Nutrition*, 24(1):215–223. https://doi.org/10.1111/anu.12549.
304. Lin, Y.H., Lin, S.M. and Shiau, S.Y. (2008). Dietary manganese requirements of juvenile tilapia, *Oreochromis niloticus* × *O. aureus*. *Aquaculture*, 284(1–4):207–210. https://doi.org/10.1016/j.aquaculture.2008.07.049.
305. Fa-yi, L. and Lawrence, A.L. (1997). Dietary manganese requirement of *Penaeus vannamei*. *Chinese Journal of Oceanology and Limnology*, 15(2):163–167. https://doi.org/10.1007/BF02850687.
306. Tan, X.Y., Xie, P., Luo, Z., Lin, H.Z., Zhao, Y.H. and Xi, W.Q. (2012). Dietary manganese requirement of juvenile yellow catfish *Pelteobagrus fulvidraco*, and effects on whole body mineral composition and hepatic intermediary metabolism. *Aquaculture*, 326:68–73. https://doi.org/10.1016/j.aquaculture.2011.11.013.
307. Zhang, H., Sun, R., Xu, W., Zhou, H., Zhang, W. and Mai, K. (2016). Dietary manganese requirement of juvenile large yellow croaker *Larimichthys crocea* (Richardson, 1846). *Aquaculture*, 450:74–79. https://doi.org/10.1016/j.aquaculture.2015.07.010.
308. Musharraf, M. and Khan, M.A. (2020). Dietary calcium requirement of fingerling Indian major carp, *Labeo rohita* (Hamilton) based on growth performance, tissue mineralization, whole body, and serum biochemical composition. *Aquaculture International*:1–15. https://doi.org/10.1007/s10499–020–00515–2.
309. Poston, H.A., Combs Jr, G.F. and Leibovitz, L. (1976). Vitamin E and selenium interrelations in the diet of Atlantic salmon (*Salmo salar*): gross, histological and biochemical deficiency signs. *Journal of Nutrition*, 106(7):892–904. https://doi.org/10.1093/jn/106.7.892.
310. Gatlin 3rd, D.M. and Wilson, R.P. (1984). Dietary selenium requirement of fingerling channel catfish. *Journal of Nutrition*, 114(3):627–633. https://doi.org/10.1093/jn/114.3.627.
311. Liu, K., Wang, X.J., Ai, Q., Mai, K. and Zhang, W. (2010). Dietary selenium requirement for juvenile cobia, *Rachycentron canadum* L. *Aquaculture Research*, 41(10):e594-e601. https://doi.org/10.1111/j.1365–2109.2010.02562.x.

312. Zhu, L., Han, D., Zhu, X., Yang, Y., Jin, J., Liu, H. and Xie, S. (2017). Dietary selenium requirement for on-growing gibel carp (*Carassius auratus gibelio* var. *CAS III*). *Aquaculture Research*, 48(6):2841–2851. https://doi.org/10.1111/are.13118.
313. Tan, F., Liang, M., Zheng, K. and Xu, H. (2015). The requirement of dietary selenium at the middle growth stage of Japanese sea bass (*Lateolabrax japonicus*). Progress in Fishery Science. *Yuye Kexue Jinzhan*, 36(3):93–100.
314. Lin, Y.H. and Shiau, S.Y. (2009). Mutual sparing of dietary requirements for alpha-tocopherol and selenium in grouper, *Epinephelus malabaricus*. *Aquaculture*, 294(3–4):242–245. https://doi.org/10.1016/j.aquaculture.2009.06.002.
315. Lin, Y.H. and Shiau, S.Y. (2005). Dietary selenium requirements of juvenile grouper, *Epinephelus malabaricus*. *Aquaculture*, 250(1–2): 356–363. https://doi.org/10.1016/j.aquaculture.2005.03.022.
316. Ning, L., Tan, Y., Wang, W., Wu, S., Chen, F., Zhang, H. and Pan, Q. (2020). Optimum selenium requirement of juvenile Nile tilapia, *Oreochromis niloticus*. *Aquaculture Nutrition*, 26(2): 528–535.
317. Lee, S., Nambi, R.W., Won, S., Katya, K. and Bai, S.C. (2016). Dietary selenium requirement and toxicity levels in juvenile Nile tilapia, *Oreochromis niloticus*. *Aquaculture*, 464:153–158. https://doi.org/10.1016/j.aquaculture.2016.06.027.
318. Hilton, J.W., Hodson, P.V. and Slinger, S.J. (1980). The requirement and toxicity of selenium in rainbow trout (*Salmo gairdneri*). *Journal of Nutrition*, 110(12):2527–2535. https://doi.org/10.1093/jn/110.12.2527.
319. Cao, J.J., Zhang, W.B., Xu, W., Mai, K.S. and Sun, R.J. (2015). Dietary selenium requirement of juvenile large yellow croaker *Larimichthys croceus*. *Acta Hydrobiologica Sinica*, 39(2): 241–249.
320. Cui, P. (2018). Effects of chromium methionine on activities of glyco-metabolism-related enzymes and expression of IR, GLUT2 and SGLT genes in common carp Cyprinus carpio. *Dalian Hai Yang da Xue Xue Bao*, Dalian, 33(3):316–322.
321. Liu, T., Wen, H., Jiang, M., Yuan, D., Gao, P., Zhao, Y., Wu, F. and Liu, W. (2010). Effect of dietary chromium picolinate on growth performance and blood parameters in grass carp fingerling, *Ctenopharyngodon idellus*. *Fish Physiology and*

Biochemistry, 36(3):565–572. https://doi.org/10.1007/s10695-009-9327-5.

322. Giri, A.K., Sahu, N.P., Saharan, N. and Dash, G. (2014). Effect of dietary supplementation of chromium on growth and biochemical parameters of *Labeo rohita* (Hamilton) fingerlings. *Indian Journal of Fisheries*, 61(2):73–81.
323. Li, H., Meng, X., Wan, W., Liu, H., Sun, M., Wang, H. and Wang, J. (2018). Effects of chromium picolinate supplementation on growth, body composition, and biochemical parameters in Nile tilapia *Oreochromis niloticus. Fish Physiology and Biochemistry*, 44(5):1265–1274.
324. Gatta, P.P., Thompson, K.D., Smullen, R., Piva, A., Testi, S. and Adams, A. (2001). Dietary organic chromium supplementation and its effect on the immune response of rainbow trout (*Oncorhynchus mykiss*). *Fish and Shellfish Immunology*, 11(5):371–382. https://doi.org/10.1006/fsim.2000.0323.
325. Tacon, A.G.J. and Beveridge, M.M. (1982). Effects of dietary trivalent chromium on rainbow trout. *Nutrition Reports International*, 25(1):49–56.
326. Lin, Y.H., Wu, J.Y. and Shiau, S.Y. (2010). Dietary cobalt can promote gastrointestinal bacterial production of vitamin B12 in sufficient amounts to supply growth requirements of grouper, *Epinephelus malabaricus. Aquaculture*, 302(1–2):89–93. https://doi.org/10.1016/j.aquaculture.2010.02.008.
327. Yatoo, M.I., Saxena, A., Deepa, P.M., Habeab, B.P., Devi, S., Jatav, R.S. and Dimri, U. (2013). Role of trace elements in animals: a review. *Veterinary World*, 6(12):963–967. https://doi.org/10.14202/vetworld.2013.963-967.
328. Hossain, M.A. and Yoshimatsu, T. (2014). Dietary calcium requirement in fishes. *Aquaculture Nutrition*, 20(1):1–11. https://doi.org/10.1111/anu.12135.
329. Davis, D.A., Lawrence, A.L. and Gatlin 3rd, D.M. (1993). Response of *Penaeus vannamei* to dietary calcium, phosphorus and calcium: phosphorus ratio. *Journal of the World Aquaculture Society*, 24(4): 504–515. https://doi.org/10.1111/j.1749-7345.1993.tb00581.x.
330. Taveekijakarn, P., Miyazaki, T., Matsumoto, M. and Arai, S. (1994). Vitamin A deficiency in cherry salmon. *Journal of Aquatic Animal Health*, 6(3):251–259. https://doi.org/10.1577/1548-8667(1994)006<0251:VADICS>2.3.CO;2.

331. Chen, S. and Li, A. (1994). Investigation on nutrition of vitamin A for shrimp *Penaeus chinensis*: 1. Effects of vitamin A on shrimp's growth and visual organ. *Dong Wu Xue Bao* (*Acta Zoologica Sinica*), 40(1):266–273.
332. Liang, M. and Ji, W. (1999). Study on nutritional requirement of vitamin A for *Penaeus chinensis* larva. *Oceanol. Limnol. Sin./Haiyang yŭ Huzhao*, 30(2):150–154.
333. Shiau, S.Y. and Chen, Y. (2000). Estimation of the dietary vitamin A requirement of juvenile grass shrimp, *Penaeus monodon*. *Journal of Nutrition*, 130(1):90–94.
334. Mohamed, J.S., Sivaram, V., Roy, T.C., Marian, M.P., Murugadass, S. and Hussain, M.R. (2003). Dietary vitamin A requirement of juvenile greasy grouper (*Epinephelus tauvina*). *Aquaculture*, 219(1–4): 693–701. https://doi.org/10.1016/S0044–8486(02)00665–8.
335. Hernandez, L.H.H., Teshima, S.I., Ishikawa, M., Alam, S., Koshio, S. and Tanaka, Y. (2005). Dietary vitamin A requirements of juvenile Japanese flounder *Paralichthys olivaceus*. *Aquaculture Nutrition*, 11(1):3–9. https://doi.org/10.1111/j.1365–2095.2004.00317.x.
336. Yang, Q. and Zhou, X. (2005). Effects of dietary vitamin A deficiency on growth performance and immune responses of juvenile Jian carp (*Cyprinus carpio* var. *Jian*). *Journal of Fishery Sciences of China*, 12(1):62–67.
337. Yang, Q.H., Zhou, X.Q., Jiang, J. and Liu, Y. (2008). Effect of dietary vitamin A deficiency on growth performance, feed utilization and immune responses of juvenile Jian carp (*Cyprinus carpio* var. *Jian*). *Aquaculture Research*, 39(8):902–906. https://doi.org/10.1111/j.1365–2109.2008.01945.x.
338. Liu, B., Zhao, Z., Brown, P.B., Cui, H., Xie, J., Habte-Tsion, H.M. and Ge, X. (2016). Dietary vitamin A requirement of juvenile Wuchang bream (*Megalobrama amblycephala*) determined by growth and disease resistance. *Aquaculture*, 450:23–30. https://doi.org/10.1016/j.aquaculture.2015.06.042.
339. Siqing, C. and Aijie, L. (1995). Studies on the nutritional requirement of vitamin D for shrimp. *Oceanologia et Limnologia Sinica*, 26(1):42–47.
340. Darias, M.J., Mazurais, D., Koumoundouros, G., Cahu, C.L. and Zambonino-Infante, J.L. (2011). Overview of vitamin D and C requirements in fish and their influence on the skeletal system. *Aquaculture*, 315(1–2):49–60. https://doi.org/10.1016/j.aquaculture.2010.12.030.

341. Shiau, S.Y. and Hwang, J.Y. (1993). Vitamin D requirements of juvenile hybrid tilapia. O. niloticus × O. aureus. *Nippon Suisan Gakkaishi*, 59(3):553–558. https://doi.org/10.2331/suisan.59.553.

342. Ling-hong, M., Xian-ping, G., Jun, X., Bo, L., Ke-bao, W., Jian, Z., Ming-chun, R., Qun-lan, Z., Liang-kun, P. and Ru-li, C. (2015). Dietary vitamin D3 requirement of Wuchang bream (*Megalobrama amblycephala*). *Aquaculture*, 436:104–109. https://doi.org/10.1016/j.aquaculture.2014.10.049.

343. Lovell, R.T. and Li, Y.P. (1978). Essentiality of vitamin D in diets of Channel catfish (*Ictalurus punctatus*). *Transactions of the American Fisheries Society*, 107(6):809–811. https://doi.org/10.1577/1548–8659(1978)107<809:EOVDID>2.0.CO;2.

344. George, J.C., Barnett, B.J., Cho, C.Y. and Slinger, S.J. (1981). Vitamin D3 and muscle function in the rainbow trout. *Cytobios*, 31:7–18.

345. Barrett, B.J., Jones, G., Cho, C.Y. and Slinger, S.J. (1982). The biological activity of 25-hydroxycholecalciferol and 1,25-dehydroxycholecalciferol for rainbow trout (*Salmo gardneri*). *Journal of Nutrition*, 112(11):2020–2026.

346. Lall, S.P., Olivier, G., Hines, J.A. and Ferguson, H.W. (1988). The role of vitamin E in nutrition and immune response of Atlantic salmon (*Salmo salar*). *Boll. Aquaculture Assoc. Canada*, 88:276–278.

347. Waagbø, R. (1994). The impact of nutritional factors on the immune system in Atlantic salmon, *Salmo salar* L.: a review. *Aquaculture Research*, 25(2):175–197. https://doi.org/10.1111/j.1365–2109.1994.tb00573.x.

348. Poston, H.A., Combs Jr, G.F. and Leibovitz, L. (1976). Vitamin E and selenium interrelations in the diet of Atlantic salmon (*Salmo salar*): gross, histological and biochemical deficiency signs. *Journal of Nutrition*, 106(7):892–904. https://doi.org/10.1093/jn/106.7.892.

349. Bowater, R.O. and Burren, B. (2007). Short communication: lateral muscular myopathy associated with vitamin E deficiency in farmed barramundi, *Lates calcarifer* (Bloch). *Journal of Fish Diseases*, 30(2):117–121. https://doi.org/10.1111/j.1365–2761.2007.00775.x.

350. Amlashi, A.S., Falahatkar, B. and Sharifi, S.D. (2012). Dietary vitamin E requirements and growth performance of young-of-the-year beluga, *Huso huso* (L.) (Chondrostei: Acipenseridae). *Fisheries & Aquatic Life*, 20(4):299–306. https://doi.org/10.2478/v10086–012–0034-y.

351. Roem, A.J., Kohler, C.C. and Stickney, R.R. (1990). Vitamin E requirements of the blue tilapia, *Oreochromis aureus* (Steindachner), in relation to dietary lipid level. *Aquaculture*, 87(2):155–164. https://doi.org/10.1016/0044–8486(90)90272-O.
352. Taveekijakarn, P., Miyazaki, T., Matsumoto, M. and Arai, S. (1996). Study on vitamin E deficiency in amago salmon. Mie Univ., Tsu, Japan, Fac. Bioresour. *Bulletin of the Faculty of Bioresources, Mie University. Tsu Iss.*, 16:17–24.
353. Wang, K., Zhou, Z. and Geng, Y. (2007). Study on the pathology of vitamin E deficiency in common carp (*Cyprinus carpio*). *Acta Hydrobiologica Sinica*, 31(3):354–362.
354. Lee, M.H. and Shiau, S.Y. (2004). Vitamin E requirements of juvenile grass shrimp, *Penaeus monodon*, and effects on non-specific immune responses. *Fish and Shellfish Immunology*, 16(4):475–485. https://doi.org/10.1016/j.fsi.2003.08.005.
355. Zhang, Y., Peng, S., Gao, Q. and Zhang, C. (2015). Effects of dietary vitamin E contents on growth, nutrition and immunity in juvenile *Epinehelus moara*. Marine fisheries/Haiyang Yuye, 37(2): 156–163.
356. Kocabas, A.M. and Dmiii, G. (1999). Dietary vitamin E requirement of hybrid striped bass (*Morone chrysops* female x *M. saxatilis* male). *Aquaculture Nutrition*, 5(1):3–7. https://doi.org/10.1046/j.1365–2095.1999.00074.x.
357. Paul, B.N., Sarkar, S. and Mohanty, S.N. (2004). Dietary vitamin E requirement of mrigal, *Cirrhinus mrigala* fry. *Aquaculture*, 242(1–4): 529–536. https://doi.org/10.1016/j.aquaculture.2004.08.037.
358. Li, Y., Huang, Y., Zhang, M., Chen, Q., Fan, W. and Zhao, Y. (2019). Effect of dietary vitamin E on growth, immunity and regulation of hepatopancreas nutrition in male oriental river prawn, *Macrobrachium nipponense*. *Aquaculture Research*, 50(7):1741–1751. https://doi.org/10.1111/are.14070.
359. Cowey, C.B., Degener, E., Tacon, A.G.J., Youngson, A. and Bell, J.G. (1984). The effect of vitamin E and oxidized fish oil on the nutrition of rainbow trout (*Salmo gairdneri*) grown at natural, varying water temperatures. *British Journal of Nutrition*, 51(3):443–451. https://doi.org/10.1079/BJN19840050.
360. Frischknecht, R., Wahli, T. and Meier, W. (1994). Comparison of pathological changes due to deficiency of vitamin C, vitamin E and combinations of vitamins C and E in rainbow trout, *Oncorhynchus*

mykiss Walbaum. *Journal of Fish Diseases*, 17(1):31–45. https://doi.org/10.1111/j.1365–2761.1994.tb00343.x.

361. Peng, L.I. and Gatlin 3rd, D.M. (2009). Dietary vitamin E requirement of the red drum *Sciaenops ocellatus*. *Aquaculture Nutrition*, 15(3): 313–319. https://doi.org/10.1111/j.1365–2095.2008.00596.x.
362. Sau, S.K., Paul, B.N., Mohanta, K.N. and Mohanty, S.N. (2004). Dietary vitamin E requirement, fish performance and carcass composition of rohu (*Labeo rohita*) fry. *Aquaculture*, 240(1–4):359–368. https://doi.org/10.1016/j.aquaculture.2004.02.008.
363. Mohseni, M. (2015). Effects of vitamin E deficiency on dietary vitamin C requirement in Siberain sturgeon (*Acipenser baerii*). *ISFJ*, 24(3):45–57.
364. Abdel-Hameid, N.A.H., Abidi, S.F. and Khan, M.A. (2012). Dietary vitamin E requirement for maximizing the growth, conversion efficiency, biochemical composition and haematological status of fingerling *Channa punctatus*. *Aquaculture Research*, 43(2): 226–238. https://doi.org/10.1111/j.1365–2109.2011.02819.x.
365. He, H. and Lawrence, A.L. (1993). Vitamin E requirement of *Penaeus vannamei*. *Aquaculture*, 118(3–4):245–255. https://doi.org/10.1016/0044–8486(93)90460-G.
366. Le, K.T., Fotedar, R. and Partridge, G. (2014). Selenium and vitamin E interaction in the nutrition of yellowtail kingfish (*Seriola lalandi*): physiological and immune responses. *Aquaculture Nutrition*, 20(3):303–313. https://doi.org/10.1111/anu.12079.
367. Li, Z.M., Hu, X.D., Wu, Q.S. and Xie, L. (2010). Recent research in vitamin E nutrition of shrimps. *Journal of Jimei University (Natural Science)*, 15(2):103–108.
368. Shiau, S.Y. and Liu, J.S. (1994). Estimation of the dietary vitamin K requirement of juvenile *Penaeus chinensis* using menadione. *Aquaculture*, 126(1–2):129–135. https://doi.org/10.1016/0044–8486(94)90254–2.
369. Krossøy, C., Waagbø, R. and Ørnsrud, R. (2011). Vitamin K in fish nutrition. *Aquaculture Nutrition*, 17(6):585–594. https://doi.org/10.1111/j.1365–2095.2011.00904.x.
370. Roy, P.K. and Lall, S.P. (2007). Vitamin K deficiency inhibits mineralization and enhances deformity in vertebrae of haddock (*Melanogrammus aeglefinus* L.). *Comparative Biochemistry and Physiology. Part B: Biochemistry and Molecular Biology*, 148(2): 174–183. https://doi.org/10.1016/j.cbpb.2007.05.006.

371. Fisher, J.P., Spitsbergen, J.M., Iamonte, T., Little, E.E. and Delonay, A. (1995). Pathological and behavioral manifestations of the "Cayuga syndrome," a thiamine deficiency in larval landlocked Atlantic salmon. *Journal of Aquatic Animal Health*, 7(4):269–283.
372. Hashimoto, Y., Arai, S. and Nose, T. (1970). Thiamine deficiency symptoms experimentally induced in the eel. *Bulletin of the Japanese Society of Scientific Fisheries*, 36(8):791–797. https://doi.org/10.2331/suisan.36.791.
373. Fitzsimons, J.D., Brown, S.B., Williston, B., Williston, G., Brown, L.R., Moore, K., Honeyfield, D.C. and Tillitt, D.E. (2009). Influence of thiamine deficiency on lake trout larval growth, foraging, and predator avoidance. *Journal of Aquatic Animal Health*, 21(4):302–314. https://doi.org/10.1577/H08–019.1.
374. Honeyfield, D.C., Hinterkopf, J.P., Fitzsimons, J.D., Tillitt, D.E., Zajicek, J.L. and Brown, S.B. (2005). Development of thiamine deficiencies and early mortality syndrome in lake trout by feeding experimental and feral fish diets containing thiaminase. *Journal of Aquatic Animal Health*, 17(1):4–12. https://doi.org/10.1577/H03–078.1.
375. Deshimaru, Q. and Kuroki, K. (1979). Studies on a purifed diet for prawn. XIV. Requirement of prawn for dietary thiamine, pyridoxine, and choline chloride. Nihon Suisan Gakkai, Shi; Tokyo, 45(3):363–367. https://doi.org/10.2331/suisan.45.363.
376. Zehra, S. and Khan, M.A. (2017). Dietary thiamin and pyridoxine requirements of fingerling Indian major carp, *Cirrhinus mrigala* (Hamilton). *Aquaculture Research*, 48(9):4945–4957. https://doi.org/10.1111/are.13313.
377. Harder, A.M., Ardren, W.R., Evans, A.N., Futia, M.H., Kraft, C.E., Marsden, J.E., Richter, C.A., Rinchard, J., Tillitt, D.E. and Christie, M.R. (2018). Thiamine deficiency in fishes: causes, consequences, and potential solutions. *Reviews in Fish Biology and Fisheries*, 28(4):865–886. https://doi.org/10.1007/s11160–018–9538-x.
378. Wen, L.M., Jiang, W.D., Liu, Y., Wu, P., Zhao, J. et al. (2015). Evaluation the effect of thiamin deficiency on intestinal immunity of young grass carp (*Ctenopharyngodon idella*). *Fish and Shellfish Immunology*, 46(2):501–515. https://doi.org/10.1016/j.fsi.2015.07.001.
379. Ren, C. (2015). Dietary thiamin requirement of GIFT *Oreochromis niloticus*. *Journal of Fisheries of China*/Shuichan Xuebao, 39(4): 539–546.

380. Jiang, M., Huang, F., Zhao, Z., Wen, H., Wu, F., Liu, W., Yang, C. and Wang, W. (2014). Dietary thiamin requirement of juvenile grass carp, *Ctenopharyngodon idella. Journal of the World Aquaculture Society*, 45(4):461–468. https://doi.org/10.1111/jwas.12132.
381. Boonyaratpalin, M. and Wanakowat, J. (1993). Effect of thiamin, riboflavin, Pantothenic Acid and inositol on growth, feed efficiency and mortality of juvenile sea bass. In *Fish nutrition in practice, INRA*. S.J. Kaushik & P. Luquet (Eds.), pp. 819–828. Cedex, France.
382. Lim, C., Leamaster, B. and Brock, J.A. (1993). Riboflavin requirement of fingerling red hybrid tilapia grown in seawater. S.J. Kaushik, P. Luquet et al. (Eds.), *Fish nutrition in practice*, INRA. Cedex, France:743–752. https://doi.org/10.1111/j.1749–7345.1993.tb00573.x.
383. Hughes, S.G. (1985). Nutritional eye diseases in salmonids: a review. *Progressive Fish-Culturist*, 47(2):81–85.
384. Murai, T. and Andrews, J.W. (1978). Thiamin requirement of channel catfish fingerlings. *Journal of Nutrition*, 108(1):176–180. https://doi.org/10.1093/jn/108.1.176.
385. Kreutzmann, H.-L. and Lehmitz, R. (1976). Influence of vitamin B1 deficiency on some physiological and morphological blood characteristics of Anguilla anguilla. *Arch. Tierernaehr*, 26(4):293–303.
386. Cowey, C.B., Adron, J.W., Knox, D. and Ball, G.T. (1975). Studies on the nutrition of marine flatfish. The thiamin requirement of turbot (*Scophthalmus maximus*). *British Journal of Nutrition*, 34(3):383–390. https://doi.org/10.1017/S000711457500044X.
387. Woodbury, L.A. (1943). Vitamin B1 deficiency in hatchery reared rainbow trout. *Transactions of the American Fisheries Society*, 72(1): 30–34. https://doi.org/10.1577/1548–8659(1942)72[30:VBDIHR]2.0.CO;2.
388. Huang, J., Tian, L., Wu, X., Yang, H. and Liu, Y. (2010). Effects of dietary riboflavin levels on antioxidant defense of the juvenile grouper *Epinephelus coioides. Fish Physiology and Biochemistry*, 36(1):55–62. https://doi.org/10.1007/s10695–008–9279–1.
389. Masumoto, T., Hardy, R.W. and Casillas, E. (1987). Comparison of transketolase activity and thiamin pyrophosphate levels in erythrocytes and liver of rainbow trout (*Salmo gairdneri*) as indicators of thiamin status. *Journal of Nutrition*, 117(8):1422–1426. https://doi.org/10.1093/jn/117.8.1422.

390. Soliman, A.K. and Wilson, R.P. (1992). Water-soluble vitamin requirements of tilapia. 2. Riboflavin requirement of blue tilapia, *Oreochromis aureus*. *Aquaculture*, 104(3–4):309–314. https://doi.org/10.1016/0044–8486(92)90212–4.
391. Phillips Jr, A.M. and Tunison, A.V. (1947). The riboflavin and Pantothenic Acid requirement of brook trout. *Transactions of the American Fisheries Society*, 74(1):81–87. https://doi.org/10.1577/1548–8659(1944)74[81:TRAPAR]2.0.CO;2.
392. Serrini, G., Zhang, Z. and Wilson, R.P. (1996). Dietary riboflavin requirement of fingerling channel catfish (*Ictalurus punctatus*). *Aquaculture*, 139(3–4):285–290. https://doi.org/10.1016/0044–8486(95)01168–4.
393. Takeuchi, L., Takeuchi, T. and Ogino, C. (1980). Riboflavin requirements in carp and rainbow trout. *Bulletin of the Japanese Society of Scientific Fisheries*, 46(6):733–737. https://doi.org/10.2331/suisan.46.733.
394. Chen, L., Feng, L., Jiang, W.D., Jiang, J., Wu, P. et al. (2015) Dietary riboflavin deficiency decreases immunity and antioxidant capacity, and changes tight junction proteins and related signaling molecules mRNA expression in the gills of young grass carp (*Ctenopharyngodon idella*). *Fish and Shellfish Immunology*, 45(2): 307–320. https://doi.org/10.1016/j.fsi.2015.04.004.
395. Woodward, B. (1984). Symptoms of severe riboflavin deficiency without ocular opacity in rainbow trout (*Salmo gairdneri*). *Aquaculture*, 37(3):275–281. https://doi.org/10.1016/0044–8486(84)90160–1.
396. Barash, H., Poston, H.A. and Rumsey, G.L. (1982). Differentiation of soluble proteins in cataracts caused by deficiencies of methionine, riboflavin or zinc in diets fed to Atlantic salmon, *Salmo salar*, rainbow trout, *Salmo gairdneri*, and lake trout, *Salvelinus namaycush*. *Cornell Veterinarian*, 72(4):361–371.
397. Zehra, S. and Khan, M.A. (2018). Dietary riboflavin requirement of fingerling *Channa punctatus* (Bloch) based on growth, conversion efficiencies, protein retention, liver riboflavin storage, RNA/DNA ratio and carcass composition. *Aquaculture Nutrition*, 24(1): 269–276. https://doi.org/10.1111/anu.12555.
398. Deng, D.F. and Wilson, R.P. (2003). Dietary riboflavin requirement of juvenile sunshine bass (*Morone chrysops*♀ × *Morone saxatilis*♂). *Aquaculture*, 218(1–4):695–701. https://doi.org/10.1016/S0044–8486(02)00513–6.

399. Amezaga, M.R. and Knox, D. (1990). Riboflavin requirements in on-growing rainbow trout, *Oncorhynchus mykiss*. *Aquaculture*, 88(1):87–98. https://doi.org/10.1016/0044–8486(90)90322-E.
400. Herman, R.L. (1985). Histopathology associated with pyridoxine deficiency in Atlantic salmon (*Salmo salar*). *Aquaculture*, 46(3): 173–177. https://doi.org/10.1016/0044–8486(85)90202–9.
401. Wang, Y., Li, X., Zhang, W., Xu, W. and Liu, W. (2013). Optimal dietary pyridoxine requirement of juvenile blunt snout bream, *Megalobrama amblycephala*. *Acta Hydrobiologica Sinica*, 37(4):632–639.
402. Kissil, G.W., Cowey, C.B., Adron, J.W. and Richards, R.H. (1981). Pyridoxine requirements of the gilthead bream, *Sparus aurata*. *Aquaculture*, 23(1–4):243–255. https://doi.org/10.1016/0044–8486(81)90018–1.
403. Lim, C., Leamaster, B.R. and Brock, J.A. (1995). Pyridoxine requirement of fingerling red hybrid tilapia grown in seawater. *Journal of Applied Aquaculture*, 5(2):49–60. https://doi.org/10.1300/J028v05n02_05.
404. Mohamed, J.S. (2001). Dietary pyridoxine requirement of the Indian catfish, *Heteropneustes fossilis*. *Aquaculture*, 194(3–4): 327–335. https://doi.org/10.1016/S0044–8486(00)00510-X.
405. He, W., Zhou, X.Q., Feng, L., Jiang, J. and Liu, Y. (2009). Dietary pyridoxine requirement of juvenile Jian carp (*Cyprinus carpio* var. *Jian*). *Aquaculture Nutrition*, 15(4):402–408. https://doi.org/10.1111/j.1365–2095.2008.00604.x.
406. Huang, J.W., Tian, L.X., Du, Z.Y., Yang, H.J. and Liu, Y.J. (2005). Pyridoxine deficiency of grouper, *Epinephelus coioides*: physiological and biochemical alteration. *Fish Physiology and Biochemistry*, 31(4):331–337. https://doi.org/10.1007/s10695–005–2832–2.
407. Zehra, S. and Khan, M.A. (2017). Dietary thiamin and pyridoxine requirements of fingerling Indian major carp, *Cirrhinus mrigala* (Hamilton). *Aquaculture Research*, 48(9):4945–4957. https://doi.org/10.1111/are.13313.
408. Zehra, S. and Khan, M.A. (2018). Dietary pyridoxine requirement of fingerling *Channa punctatus* (Bloch) based on growth performance, liver pyridoxine concentration, and carcass composition. *Journal of Applied Aquaculture*, 30(3):238–255. https://doi.org/10.1080/10454438.2018.1456999.
409. Takeuchi, T., Watanabe, K., Satoh, S. and Watanabe, T. (1992). Requirement of grass carp fingerlings for alpha-t ocopherol. *Nippon*

Suisan Gakkaishi, 58(9):1743–1749. https://doi.org/10.2331/suisan.58.1743.

410. Lovell, R.T., Miyazaki, T. and Rabegnator, S. (1984). Requirement for α-tocopherol by channel catfish fed diets low in polyunsaturated triglycerides. *Journal of Nutrition*, 114(5):894–901. https://doi.org/10.1093/jn/114.5.894.
411. Watanabe, T., Takeuchi, T., Wada, M. and Uehara, M. (1981). The relationship between dietary lipid levels and alpha-tocopherol requirement of rainbow trout. *Nippon Suisan Gakkaishi*, 47(11):1463–1471. https://doi.org/10.2331/suisan.47.1463.
412. Watanabe, T., Takeuchi, T., Matsui, M., Ogino, C. and Kawabata, T. (1977). Effect of alpha-tocopherol deficiency on carp. VII. The relationship between dietary levels of linoleate and alpha-tocopherol requirement. *Nippon Suisan Gakkaishi*, 43(8):935–946. https://doi.org/10.2331/suisan.43.935.
413. Agrawal, N.K. and Mahajan, C.L. (1983). Pathology of vitamin B-6 deficiency in *Channa* (= *Ophiocephalus*) *punctatus* Bloch. *Journal of Fish Diseases*, 6(5):439–450. https://doi.org/10.1111/j.1365–2761.1983.tb00098.x.
414. Shiau, S.Y. and Wu, M.H. (2003). Dietary vitamin B6 requirement of grass shrimp, *Penaeus monodon*. *Aquaculture*, 225(1–4):397–404. https://doi.org/10.1016/S0044–8486(03)00304–1.
415. Soliman, A.K. and Wilson, R.P. (1992). Water-soluble vitamin requirements of tilapia. 1 Pantothenic Acid requirement of blue tilapia, Oreochromis aureus. *Aquaculture*, 104(1–2):121–126. https://doi.org/10.1016/0044–8486(92)90142–8.
416. Phillips Jr, A.M. and Tunison, A.V. (1947). The riboflavin and Pantothenic Acid requirement of brook trout. *Transactions of the American Fisheries Society*, 74(1):81–87. https://doi.org/10.1577/1548–8659(1944)74[81:TRAPAR]2.0.CO;2.
417. Wilson, R.P., Bowser, P.R. and Poe, W.E. (1983). Dietary Pantothenic Acid requirement of fingerling channel catfish. *Journal of Nutrition*, 113(10):2124–2128. https://doi.org/10.1093/jn/113.10.2124.
418. Li, L., Feng, L., Jiang, W.D., Jiang, J., Wu, P., Kuang, S.Y., Tang, L., Tang, W.N., Zhang, Y.A., Zhou, X.Q. and Liu, Y. (2015). Dietary Pantothenic Acid deficiency and excess depress the growth, intestinal mucosal immune and physical functions by regulating NF-κB, TOR, Nrf2 and MLCK signaling pathways in grass

carp (*Ctenopharyngodon idella*). *Fish and Shellfish Immunology*, 45(2):399–413. https://doi.org/10.1016/j.fsi.2015.04.030.

419. Xun, P., Lin, H., Huang, Z., Zhou, C., Yang, Y.; et al. (2018). Dietary Pantothenic Acid requirement of juvenile golden pompano (*Trachinotus ovatus*). *Nanfang Shuichan Kexue = South China Fisheries Science*. Iss, Guangzhou:5.
420. Raggi, T., Buentello, A. and Gatlin 3rd, D.M. (2016). Characterization of Pantothenic Acid deficiency and the dietary requirement of juvenile hybrid striped bass, *Morone chrysops × M. saxatilis*. *Aquaculture*, 451:326–329. https://doi.org/10.1016/j.aquaculture.2015.09.028.
421. Poston, H.A. and Page, J.W. (1982). Gross and histological signs of dietary deficiencies of biotin and Pantothenic Acid in lake trout, *Salvelinus namaycush*. *Cornell Veterinarian*, 72(3):242–261.
422. Lin, Y.H., Lin, H.Y. and Shiau, S.Y. (2012). Estimation of dietary Pantothenic Acid requirement of grouper, *Epinephelus malabaricus* according to physiological and biochemical parameters. *Aquaculture*, 324:92–96. https://doi.org/10.1016/j.aquaculture.2011.10.020.
423. Zehra, S. and Khan, M.A. (2018). Dietary Pantothenic Acid requirement of fingerling *Channa punctatus* (Bloch) based on growth, feed conversion, liver Pantothenic Acid concentration and carcass composition. *Aquaculture Nutrition*, 24(5):1436–1443. https://doi.org/10.1111/anu.12680.
424. Chávez-Sánchez, M.C., Pimentel-Acosta, C.A., Olvera-Novoa, M.A., Hernández, C., Río-Zaragoza, O.B.D., Abad-Rosales, S.M. and Martínez-Rodríguez, I. (2018). The Pantothenic Acid requirement in juvenile spotted rose snapper *Lutjanus guttatus* (Steindachner, 1869). *Latin American Journal of Aquatic Research*, 46(5): 1001–1010. https://doi.org/10.3856/vol46-issue5-fulltext-13.
425. Li, X.F., Wang, T.J., Qian, Y., Jiang, G.Z., Zhang, D.D. and Liu, W.B. (2017). Dietary niacin requirement of juvenile blunt snout bream *Megalobrama amblycephala* based on a dose-response study. *Aquaculture Nutrition*, 23(6):1410–1417.
426. Li, S.Q., Feng, L., Jiang, W.D., Liu, Y., Wu, P. et al. (2016). Deficiency of dietary niacin decreases digestion and absorption capacities via declining the digestive and brush border enzyme activities and downregulating those enzyme gene transcription related to TOR pathway of the hepatopancreas and intestine in young

grass carp (*Ctenopharyngodon idella*). *Aquaculture Nutrition*, 22(6): 1267–1282. https://doi.org/10.1111/anu.12333.

427. Li, S.Q., Feng, L., Jiang, W.D., Liu, Y., Jiang, J., Wu, P., Kuang, S.Y., Tang, L., Tang, W.N., Zhang, Y.A. and Zhou, X.Q. (2016). Deficiency of dietary niacin impaired gill immunity and antioxidant capacity, and changes its tight junction proteins via regulating NF-κB, TOR, Nrf2 and MLCK signaling pathways in young grass carp (*Ctenopharyngodon idella*). *Fish and Shellfish Immunology*, 55:212–222. https://doi.org/10.1016/j.fsi.2016.05.005.
428. Feng, L., Li, S.Q., Jiang, W.D., Liu, Y., Jiang, J. et al. (2016). Deficiency of dietary niacin impaired intestinal mucosal immune function via regulating intestinal NF-κB, Nrf2 and MLCK signaling pathways in young grass carp (*Ctenopharyngodon idella*). *Fish and Shellfish Immunology*, 49:177–193. https://doi.org/10.1016/j.fsi.2015.12.015.
429. Shiau, S.Y. and Suen, G.S. (1994). The dietary requirement of juvenile grass shrimp (*Penaeus monodon*) for niacin. *Aquaculture*, 125(1–2): 139–145. https://doi.org/10.1016/0044–8486(94)90290–9.
430. Mohamed, J.S. and Ibrahim, A. (2001). Quantifying the dietary niacin requirement of the Indian catfish, *Heteropneustes fossilis* (Bloch), fingerlings. *Aquaculture Research*, 32(3):157–162. https://doi.org/10.1046/j.1365–2109.2001.00530.x.
431. Jiang, M., Huang, F., Wen, H., Yang, C., Wu, F., Liu, W. and Tian, J. (2014). Dietary niacin requirement of GIFT tilapia, *Oreochromis niloticus*, reared in freshwater. *Journal of the World Aquaculture Society*, 45(3):333–341. https://doi.org/10.1111/jwas.12119.
432. Poston, H.A. and Wolfe, M.J. (1985). Niacin requirement for optimum growth, feed conversion and protection of rainbow trout, *Salmo gairdneri* Richardson, from ultraviolet-B irradiation. *Journal of Fish Diseases*, 8(5):451–460. https://doi.org/10.1111/j.1365–2761.1985.tb01278.x.
433. Zehra, S. and Khan, M.A. (2018). Dietary niacin requirement of fingerling *Channa punctatus* (Bloch). *Journal of Applied Ichthyology*, 34(4):929–936. https://doi.org/10.1111/jai.13604.
434. Shi, L., Feng, L., Jiang, W.D., Liu, Y., Jiang, J., Wu, P., Kuang, S.Y., Tang, L., Tang, W.N., Zhang, Y.A. and Zhou, X.Q. (2016). Immunity decreases, antioxidant system damages and tight junction changes in the intestine of grass carp (*Ctenopharyngodon idella*) during folic acid deficiency: regulation of NF-κB, Nrf2 and MLCK

mRNA levels. *Fish and Shellfish Immunology*, 51:405–419. https://doi.org/10.1016/j.fsi.2016.02.029.

435. Shi, L., Feng, L., Jiang, W.D., Liu, Y., Jiang, J. et al. (2015). Folic acid deficiency impairs the gill health status associated with the NF-κB, MLCK and Nrf2 signaling pathways in the gills of young grass carp (*Ctenopharyngodon idella*). *Fish and Shellfish Immunology*, 47(1):289–301. https://doi.org/10.1016/j.fsi.2015.09.023.
436. Shiau, S.Y. and Huang, S.Y. (2001). Dietary folic acid requirement determined for grass shrimp, *Penaeus monodon*. *Aquaculture*, 200(3–4): 339–347. https://doi.org/10.1016/S0044–8486(00)00598–6.
437. Lin, Y.H., Lin, H.Y. and Shiau, S.Y. (2011). Dietary folic acid requirement of grouper, *Epinephelus malabaricus*, and its effects on non-specific immune responses. *Aquaculture*, 317(1–4):133–137. https://doi.org/10.1016/j.aquaculture.2011.04.010.
438. Wu, J.P., Wu, F., Jiang, M., Wen, H., Wei, Q.W., Liu, W., Tian, J., Huang, F. and Yang, C.G. (2016). Dietary folic acid (FA) requirement of a genetically improved Nile tilapia *Oreochromis niloticus* (Linnaeus, 1758). *Journal of Applied Ichthyology*, 32(6):1155–1160. https://doi.org/10.1111/jai.13151.
439. Qian, Y., Li, X.F., Sun, C.X., Jiang, G.Z., Xu, W.N., Wang, Y. and Liu, W.B. (2014). Dietary biotin requirement of juvenile blunt snout bream, *Megalobrama amblycephala*. *Aquaculture Nutrition*, 20(6):616–622. https://doi.org/10.1111/anu.12114.
440. Khan, Y.M. and Khan, M.A. (2019). Dietary biotin requirement of fingerling *catla catla* (Hamilton) based on growth, feed conversion efficiency, and liver biotin concentration. *Journal of the World Aquaculture Society*, 50(3):674–683. https://doi.org/10.1111/jwas.12529.
441. Shaik Mohamed, J. (2001). Dietary biotin requirement determined for Indian catfish, *Heteropneustes fossilis* (Bloch), fingerlings. *Aquaculture Research*, 32(9):709–716. https://doi.org/10.1046/j.1365–2109.2001.00606.x.
442. Li, J., Zhang, L., Mai, K., Ai, Q., Wan, J., Zhang, C. and Zhang, J. (2010). Estimation of dietary biotin requirement of Japanese sea bass, *Lateolabrax japonicus C*. *Aquaculture Nutrition*, 16(3): 231–236. https://doi.org/10.1111/j.1365–2095.2009.00655.x.
443. Sarker, P.K., Yossa, R., Karanth, S., Ekker, M. and Vandenberg, G.W. (2012). Influences of dietary biotin and avidin on growth, survival, deficiency syndrome and hepatic gene expression of juvenile

Nile tilapia *Oreochromis niloticus*. *Fish Physiology and Biochemistry*, 38(4):1183–1193. https://doi.org/10.1007/s10695–012–9604–6.
444. Shiau, S.Y. and Chin, Y.H. (1999). Estimation of the dietary biotin requirement of juvenile hybrid tilapia, *Oreochromis niloticus* × *O. aureus*. *Aquaculture*, 170(1):71–78. https://doi.org/10.1016/S0044–8486(98)00391–3.
445. Walton, M.J., Cowey, C.B. and Adron, J.W. (1984). Effects of biotin deficiency in rainbow trout (*Salmo gairdneri*) fed diets of different lipid and carbohydrate content. *Aquaculture*, 37(1):21–38. https://doi.org/10.1016/0044–8486(84)90041–3.
446. Zehra, S. and Khan, M.A. (2019). Dietary biotin requirement of fingerling *Channa punctatus* (Bloch). *Journal of Applied Aquaculture*, 31(3):236–253. https://doi.org/10.1080/10454438.2018.1545722.
447. Mohamed, J.S., Ravisankar, B. and Ibrahim, A. (2000). Quantifying the dietary biotin requirement of the catfish, *Clarias batrachus*. *Aquaculture International*, 8(1):9–18.
448. Li, X.F., Wang, F., Qian, Y., Jiang, G.Z., Zhang, D.D. and Liu, W.B. (2016). Dietary vitamin B12 requirement of fingerling blunt snout bream *Megalobrama amblycephala* determined by growth performance, digestive and absorptive capability and status of the GH-IGF-I axis. *Aquaculture*, 464:647–653. https://doi.org/10.1016/j.aquaculture.2016.08.019.
449. Shiau, S.-Y. and Lung, C.-Q. (1993). Estimation of the vitamin B12 requirement of the grass shrimp, *Penaeus monodon*. *Aquaculture*, 117(1–2):157–163. https://doi.org/10.1016/0044–8486(93)90132-I.
450. Hansen, A.K.G., Kortner, T.M., Krasnov, A., Björkhem, I., Penn, M. and Krogdahl, Å. (2020). Choline supplementation prevents diet induced gut mucosa lipid accumulation in post-smolt Atlantic salmon (*Salmo salar* L.). *BMC Veterinary Research*, 16(1):32. https://doi.org/10.1186/s12917–020–2252–7.
451. Hansen, A.K., Kortner, T.M., Denstadli, V., Måsøval, K., Björkhem, I., Grav, H.J. and Krogdahl, Å. (2020). Dose-response relationship between dietary choline and lipid accumulation in pyloric enterocytes of Atlantic salmon (*Salmo salar* L.) in seawater. *British Journal of Nutrition*, 123(10):1081–1093. https://doi.org/10.1017/S0007114520000434.
452. Jiang, G.Z., Wang, M., Liu, W.B., Li, G.F. and Qian, Y. (2013). Dietary choline requirement for juvenile blunt snout bream,

Megalobrama amblycephala. Aquaculture Nutrition, 19(4):499–505. https://doi.org/10.1111/anu.12001.

453. Mai, K., Xiao, L., Ai, Q., Wang, X., Xu, W., Zhang, W., Liufu, Z. and Ren, M. (2009). Dietary choline requirement for juvenile cobia, *Rachycentron canadum. Aquaculture*, 289(1–2):124–128. https://doi.org/10.1016/j.aquaculture.2009.01.016.
454. Liu, T., Li, A. and Zhang, J. (1993). Studies on vitamin nutrition for the shrimp *Penaeus chinensis*. 10. Studies on the choline chloride and inositol requirements in the shrimp *Penaeus chinensis. Journal of Ocean University of Qingdao*. Qingdao Haiyang Daxue Xueba, Qingdao, 23(4):67–74.
455. Duan, Y., Zhu, X., Han, D., Yang, Y. and Xie, S. (2012). Dietary choline requirement in slight methionine-deficient diet for juvenile gibel carp (*Carassius auratus gibelio*). *Aquaculture Nutrition*, 18(6):620–627. https://doi.org/10.1111/j.1365–2095.2011.00930.x.
456. Wu, P., Jiang, W.D., Jiang, J., Zhao, J., Liu, Y., Zhang, Y.A., Zhou, X.Q. and Feng, L. (2016). Dietary choline deficiency and excess induced intestinal inflammation and alteration of intestinal tight junction protein transcription potentially by modulating NF-κB, STAT and p38 MAPK signaling molecules in juvenile Jian carp. *Fish and Shellfish Immunology*, 58:462–473. https://doi.org/10.1016/j.fsi.2016.09.055.
457. Shiau, S.Y. and Lo, P.S. (2000). Dietary choline requirements of juvenile hybrid tilapia, *Oreochromis niloticus* × *O. aureus. Journal of Nutrition*, 130(1):100–103. https://doi.org/10.1093/jn/130.1.100.
458. Won, S., Lee, S., Hamidoghli, A., Lee, S. and Bai, S.C. (2019). Dietary choline requirement of juvenile olive flounder (*Paralichthys olivaceus*). *Aquaculture Nutrition*, 25(6):1281–1288. https://doi.org/10.1111/anu.12948.
459. Craig, S.R. and Gatlin 3rd, D.M. (1996). Dietary choline requirement of juvenile red drum (*Sciaenops ocellatus*). *Journal of Nutrition*, 126(6):1696–1700. https://doi.org/10.1093/jn/126.6.1696.
460. Gong, H., Lawrence, A.L., Jiang, D.H. and Gatlin 3rd, D.M. (2003). Effect of dietary phospholipids on the choline requirement of *Litopenaeus vannamei* juveniles. *Journal of the World Aquaculture Society*, 34(3):289–299. https://doi.org/10.1111/j.1749–7345.2003.tb00067.x.

461. Hung, S.S. (1989). Choline requirement of hatchery-produced juvenile white sturgeon (*Acipenser transmontanus*). *Aquaculture*, 78(2):183–194. https://doi.org/10.1016/0044–8486(89)90031–8.
462. Twibell, R.G. and Brown, P.B. (2000). Dietary choline requirement of juvenile yellow perch (*Perca flavescens*). *Journal of Nutrition*, 130(1):95–99. https://doi.org/10.1093/jn/130.1.95.
463. Gong, W., Lei, W., Zhu, X., Yang, Y., Han, D. and Xie, S. (2014). Dietary myo-inositol requirement for juvenile gibel carp (*Carassius auratus gibelio*). *Aquaculture Nutrition*, 20(5):514–519. https://doi.org/10.1111/anu.12104.
464. Shiau, S.Y. and Su, S.L. (2004). Dietary inositol requirement for juvenile grass shrimp, *Penaeus monodon*. *Aquaculture*, 241(1–4): 1–8. https://doi.org/10.1016/j.aquaculture.2004.01.013.
465. Li, Y., Wu, J., Chen, X., Chu, Z., Jin, J. and Tang, D. (2020). Dietary myo-inositol requirements of juvenile hybrid sturgeon (*Acipenser baerii*♀ × *A. schrenckii*♂). *Aquaculture Research*, 51(3):1143–1151. https://doi.org/10.1111/are.14461.
466. Lee, B.J., Lee, K.J., Lim, S.J. and Lee, S.M. (2008). Dietary myo-inositol requirement for Olive flounder, *Paralichthys olivaceus* (Temminch et Schlegel). *Aquaculture Research*, 40(1):83–90. https://doi.org/10.1111/j.1365–2109.2008.02067.x.
467. Duncan, P.L. and Lovell, R.T. (1994). Influence of vitamin C on the folate requirement of channel catfish, *Ictalurus punctatus*, for growth, hematopoiesis, and resistance to *Edwardsiella ictaluri* infection. *Aquaculture*, 127(2–3):233–244. https://doi.org/10.1016/0044–8486(94)90429–4.
468. Eya, J.C. (1996). 'Broken-skull disease' in African catfish *Clarias gariepinus* is related to a dietary deficiency of ascorbic acid. *Journal of the World Aquaculture Society*, 27(4):493–498. https://doi.org/10.1111/j.1749–7345.1996.tb00635.x.
469. Lewis-McCrea, L.M. and Lall, S.P. (2010). Effects of phosphorus and vitamin C deficiency, vitamin A toxicity, and lipid peroxidation on skeletal abnormalities in Atlantic halibut (*Hippoglossus hippoglossus*). *Journal of Applied Ichthyology*, 26(2):334–343. https://doi.org/10.1111/j.1439–0426.2010.01431.x.
470. Adham, K.G., Hashem, H.O., Abu-Shabana, M.B. and Kamel, A.H. (2000). Vitamin C deficiency in the catfish *Clarias gariepinus*. *Aquaculture Nutrition*, 6(2):129–139. https://doi.org/10.1046/j.1365–2095.2000.00139.x.

471. Lall, S.P., Olivier, G., Weerakoon, D.E.M. and Hines, J.A. (1989). The effect of vitamin C deficiency and excess on immune response in Atlantic salmon (*Salmo salar* L.). In *Proceedings of the Third International Symposium on Feeding and Nutrition in Fish*, August 28th–1 Sept. 1989, pp. 427–441. Tokyo University of Fisheries, Toba, Japan.
472. Hamre, K., Waagbø, R., Berge, R.K. and Lie, Ø. (1997). Vitamins C and E interact in juvenile Atlantic salmon (*Salmo salar*, L.). *Free Radical Biology and Medicine*, 22(1–2):137–149. https://doi.org/10.1016/S0891-5849(96)00281-X.
473. Sandnes, K., Torrissen, O. and Waagbø, R. (1992). The minimum dietary requirement of vitamin C in Atlantic salmon (*Salmo salar*) fry using Ca ascorbate-2-monophosphate as dietary source. *Fish Physiology and Biochemistry*, 10(4):315–319. https://doi.org/10.1007/BF00004480.
474. Xie, Z. and Niu, C. (2006). Dietary ascorbic acid requirement of juvenile Ayu (*Plecoglossus altivelis*). *Aquaculture Nutrition*, 12(2): 151–156. https://doi.org/10.1111/j.1365–2095.2006.00395.x.
475. Dabrowski, K., Moreau, R., El-Saidy, D. and Ebeling, J. (1996). Ontogenetic sensitivity of channel catfish to ascorbic acid deficiency. *Journal of Aquatic Animal Health*, 8(1):22–27. https://doi.org/10.1577/1548–8667(1996)008<0022:OSOCCT>2.3.CO;2.
476. Lim, C. and Lovell, R.T. (1978). Pathology of the vitamin C deficiency syndrome in channel catfish (*Ictalurus punctatus*). *Journal of Nutrition*, 108(7):1137–1146. https://doi.org/10.1093/jn/108.7.1137.
477. Huang, F., Wu, F., Zhang, S., Jiang, M., Liu, W., Tian, J., Yang, C. and Wen, H. (2017). Dietary vitamin C requirement of juvenile Chinese sucker (*Myxocyprinus asiaticus*). *Aquaculture Research*, 48(1):37–46. https://doi.org/10.1111/are.12858.
478. Xiao, L.D., Mai, K.S., Ai, Q.H., Xu, W., Wang, X.J., Zhang, W.B. and Liufu, Z.G. (2010). Dietary ascorbic acid requirement of cobia, *Rachycentron canadum* Linneaus. *Aquaculture Nutrition*, 16(6): 582–589. https://doi.org/10.1111/j.1365–2095.2009.00695.x.
479. Halver, J.E., Ashley, L.M. and Smith, R.R. (1969). Ascorbic acid requirements of Coho salmon and rainbow trout. *Transactions of the American Fisheries Society*, 98(4):762–771. https://doi.org/10.1577/1548–8659(1969)98[762:AAROCS]2.0.CO;2.

480. Alexis, M.N., Karanikolas, K.K. and Richards, R.H. (1997). Pathological findings owing to the lack of ascorbic acid in cultured gilthead bream (*Sparus aurata* L.). *Aquaculture*, 151(1–4):209–218. https://doi.org/10.1016/S0044–8486(96)01475–5.
481. D'Abramo, L.R., Moncreiff, C.A., Holcomb, F.P., Montanez, J.L. and Buddington, R.K. (1994). Vitamin C requirement of the juvenile freshwater prawn, *Macrobrachium rosenbergii*. *Aquaculture*, 128 (3–4):269–275. https://doi.org/10.1016/0044–8486(94)90316–6.
482. Xu, H.J., Jiang, W.D., Feng, L., Liu, Y., Wu, P., Jiang, J., Kuang, S.Y., Tang, L., Tang, W.N., Zhang, Y.A. and Zhou, X.Q. (2016). Dietary vitamin C deficiency depressed the gill physical barriers and immune barriers referring to Nrf2, apoptosis, MLCK, NF-κB and TOR signaling in grass carp (*Ctenopharyngodon idella*) under infection of Flavobacterium columnare. *Fish and Shellfish Immunology*, 58:177–192. https://doi.org/10.1016/j.fsi.2016.09.029.
483. Xu, H.J., Jiang, W.D., Feng, L., Liu, Y., Wu, P., Jiang, J., Kuang, S.Y., Tang, L., Tang, W.N., Zhang, Y.A. and Zhou, X.Q. (2016). Dietary vitamin C deficiency depresses the growth, head kidney and spleen immunity and structural integrity by regulating NF-κB, TOR, Nrf2, apoptosis and MLCK signaling in young grass carp (*Ctenopharyngodon idella*). *Fish and Shellfish Immunology*, 52: 111–138. https://doi.org/10.1016/j.fsi.2016.02.033.
484. Hsu, T.S. and Shiau, S.Y. (1998). Comparison of vitamin C requirement for maximum growth of grass shrimp, *Penaeus monodon*, with L-ascorbyl-2-monophosphate-Na and L-ascorbyl-2-monophosphate-Mg. *Aquaculture*, 163(3–4):203–213. https://doi.org/10.1016/S0044–8486(98)00230–0.
485. Shiau, S.-Y. and Hsu, T.-S. (1994). Vitamin C requirement of grass shrimp, *Penaeus monodon*, as determined with L-ascorbyl-2-monophosphate. *Aquaculture*, 122(4):347–357. https://doi.org/10.1016/0044–8486(94)90343–3.
486. Boonyaratpalin, M. and Phongmaneerat, J. (1995). Ascorbic acid derivative requirement of *Penaeus monodon*. *Phuket Marine Biological Center Research Bulletin (Thailand)*.
487. Shiau, S.-Y. and Jan, F.-L. (1992). Dietary ascorbic acid requirement of juvenile tilapia *Oreochromis niloticus* x *O. aureus*. *Nippon Suisan Gakkaishi*, 58(4):671–675. https://doi.org/10.2331/suisan.58.671.

488. Shiau, S.Y. and Hsu, T.S. (1999). Quantification of vitamin C requirement for juvenile hybrid tilapia, *Oreochromis niloticus* × *Oreochromis aureus*, with L-ascorbyl-2-monophosphate-Na and L-ascorbyl-2-monophosphate-Mg. *Aquaculture*, 175(3–4): 317–326. https://doi.org/10.1016/S0044–8486(99)00103–9.
489. Sealey, W.M. and Gatlin 3rd, D.M. (1999). Dietary vitamin C requirement of hybrid striped bass *Morone chrysops*♀ × *M. saxatilis*♂. *Journal of the World Aquaculture Society*, 30(3):297–301. https://doi.org/10.1111/j.1749–7345.1999.tb00679.x.
490. Sahoo, M. and Mukhopadhyay, P.K. (2008). Dietary ascorbic acid requirements of fingerlings of genetically improved rohu (*Labeo rohita* Ham). *Israeli Journal of Aquaculture/Bamidgeh*, 60(2): 100–106. https://doi.org/10.46989/001c.20479.
491. Moe, Y.Y., Koshio, S., Ishikawa, M., Teshima, S.I., Panganiban Jr, A., Thu, M., Michael, F.R. and Ren, T. (2005). Vitamin C requirement of kuruma shrimp postlarvae, *Marsupenaeus japonicus* (Bate), using L-ascorbyl-2-monophosphate-Na/Ca. *Aquaculture Research*, 36(8):739–745. https://doi.org/10.1111/j.1365–2109.2005.01260.x.
492. Chen, Y.J., Yuan, R.M., Liu, Y.J., Yang, H.J., Liang, G.Y. and Tian, L.X. (2015). Dietary vitamin C requirement and its effects on tissue antioxidant capacity of juvenile largemouth bass, *Micropterus salmoides*. *Aquaculture*, 435:431–436. https://doi.org/10.1016/j.aquaculture.2014.10.013.
493. Lin, M.F. and Shiau, S.Y. (2005). Requirements of vitamin C (l-ascorbyl-2-sulphate and l-ascorbyl-2-polyphosphate) and its effects on non-specific immune responses of grouper, *Epinephelus malabaricus*. *Aquaculture Nutrition*, 11(3):183–189. https://doi.org/10.1111/j.1365–2095.2005.00336.x.
494. Huang, F., Jiang, M., Wen, H., Wu, F., Liu, W., Tian, J. and Shao, H. (2016). Dietary vitamin C requirement of genetically improved farmed tilapia, *Oreochromis niloticus*. *Aquaculture Research*, 47(3):689–697. https://doi.org/10.1111/are.12527.
495. Martins, M.L. (1995). Effect of ascorbic acid deficiency on the growth, gill filament lesions and behavior of Pacu fry (*Piaractus mesopotamicus* Holmberg, 1887). *Brazilian Journal of Medical and Biological Research*, 28(5):563–568.
496. Dabrowski, K., El-Fiky, N., Köck, G., Frigg, M. and Wieser, W. (1990). Requirement and utilization of ascorbic acid and ascorbic

sulfate in juvenile rainbow trout. *Aquaculture*, 91(3–4):317–337. https://doi.org/10.1016/0044–8486(90)90197-U.

497. Akand, A.M., Sato, M., Yoshinaka, R. and Ikeda, S. (1987). Haematological changes due to dietary ascorbic acid deficiency at various levels of calcium and phosphorus in rainbow trout. *Current Science Bangalore*, 56(7):298–301.
498. Sato, M., Yoshinaka, R. and Ikeda, S. (1978). Dietary ascorbic acid requirement of rainbow trout for growth and collagen formation. Nihon Suisan Gakkai, Shi; Tokyo, 44(9):1029–1035. https://doi.org/10.2331/suisan.44.1029.
499. Falahatkar, B., Dabrowski, K. and Arslan, M. (2011). Ascorbic acid turnover in rainbow trout, Oncorhynchus mykiss: is there a vitamin enrichment effect during embryonic period on the juvenile fish "sensitivity" to deficiency? *Aquaculture*, 320(1–2):99–105. https://doi.org/10.1016/j.aquaculture.2011.08.012.
500. Aguirre, P. and Gatlin 3rd, D.M. (1999). Dietary vitamin C requirement of red drum *Sciaenops ocellatus*. *Aquaculture Nutrition*, 5(4): 247–249. https://doi.org/10.1046/j.1365–2095.1999.00114.x.
501. Chávez-Sánchez, M.C., Olvera-Novoa, M.A., Osuna-Durán, B., de la Parra, I.A., Abad-Rosales, S.M. and Martínez-Rodríguez, I. (2014). Ascorbic acid requirement and histopathological changes due to its deficiency in juvenile spotted rose snapper *Lutjanus guttatus* (Steindachner, 1869). *Aquaculture International*, 22(6): 1891–1909. https://doi.org/10.1007/s10499–014–9790–6.
502. Agrawal, N.K. and Mahajan, C.L. (1981). Effect of ascorbic acid deficiency on the uptake of iodine by thyroid and non-thyroid tissues of an air-breathing freshwater fish *Channa punctatus* Bloch. *Journal of Fish Biology*, 18(4):411–416. https://doi.org/10.1111/j.1095–8649.1981.tb03782.x.
503. Agrawal, N.K. and Mahajan, C.L. (1980). Hematological changes due to vitamin C deficiency in *Channa punctatus* Bloch. *Journal of Nutrition*, 110(11):2172–2181. https://doi.org/10.1093/jn/110.11.2172.
504. Mahajan, C.L. and Agrawal, N.K. (1979). Vitamin C deficiency in *Channa punctatus* Bloch. *Journal of Fish Biology*, 15(5):613–622. https://doi.org/10.1111/j.1095–8649.1979.tb03652.x.
505. Coustans, M.F.O. and Guillaume, J., Metailler, R., Dugornay, O. and Messager, J., Metailler, R., Dugornay, O. and Messager, J.L. (1990). Effect of an ascorbic acid deficiency on tyrosinemia and renal granulomatous disease in turbot (*Scophthalmus maximus*) interaction

with a slight polyhypovitaminosis. *Comparative Biochemistry and Physiology. A Comparative Physiology*, 97(2):145–152. https://doi.org/10.1016/0300–9629(90)90161-K.

506. Ibiyo, L.M.O., Atteh, J.O., Omotosho, J.S. and Madu, C.T. (2007). Vitamin C (ascorbic acid) requirements of *Heterobranchus longifilis* fingerlings. *African Journal of Biotechnology*, 6(13).
507. He, H. and Lawrence, A.L. (1993). Vitamin C requirements of the shrimp *Penaeus vannamei. Aquaculture*, 114(3–4):305–316. https://doi.org/10.1016/0044–8486(93)90305-I.
508. Lee, K.J. and Dabrowski, K. (2003). Interaction between vitamins C and E affects their tissue concentrations, growth, lipid oxidation, and deficiency symptoms in yellow perch (*Perca flavescens*). *British Journal of Nutrition*, 89(5):589–596. https://doi.org/10.1079/BJN2003819.
509. Macconnell, E. and Barrows, F.T. (1993). Pathological changes associated with vitamin C deficiency in walleyes. *Journal of Aquatic Animal Health*, 5(4):287–293. https://doi.org/10.1577/1548–8667(1993)005<0287:PCAWVC>2.3.CO;2.
510. Hertrampf, J.W. and Piedad-Pascual, F. (2012). *Handbook on ingredients for aquaculture feeds.* Kluwer Academic Publishers, London.
511. Anglesea, J.D. and Jackson, A.J. (1985). Thiaminase activity in fish silage and moist fish feed. *Animal Feed Science and Technology*, 13(1–2):39–46. https://doi.org/10.1016/0377–8401(85)90040–9.
512. Liener, I. ed. (1980). *Toxic constituents of plant foodstuffs.* Academic Press, New York.
513. Tacon, A.G.J. (1997). Fish-meal replacers: review of antinutrients within oilseeds and pulses – A limiting factor for the aquafeed green revolution. *Cahiers Options Méditerranéennes*, 22:153–182.
514. Vikas, K., Debtanu, B., Kundan, K., Vikash, K., Mandal, S.C. and Clercq, E.D. (2012). Anti-nutritional factors in plant feedstuffs used in aquafeeds. World. *Aquaculture*, 43(3):64–68.
515. Riaz, M.N., Asif, M. and Ali, R. (2009). Stability of vitamins during extrusion. *Critical Reviews in Food Science and Nutrition*, 49(4):361–368. https://doi.org/10.1080/10408390802067290.
516. Barrows, F.T., Gaylord, T.G., Sealey, W.M., Porter, L. and Smith, C.E. (2008). The effect of vitamin premix in extruded plant-based and fish meal based diets on growth efficiency and health of rainbow trout, *Oncorhynchus mykiss. Aquaculture*, 283(1–4):148–155. https://doi.org/10.1016/j.aquaculture.2008.07.014.

517. Gadient, M. (1986). Effect of pelleting on nutritional quality of feeds. Maryland Nutrition Conference for Feed Manufacturers, College Park, Maryland, USA, pp. 73–79.
518. Broz, J., Schai, E. and Gadient, M. (1997). Micronutrient stability in feed processing. *ASA Technical Bulletin, FT*, 42(1):8.
519. Gadient, M. and Schai, E. (1994). Leaching of various vitamins from shrimp feed. *Aquaculture*, 124(1–4):201–205. https://doi.org/10.1016/0044–8486(94)90378–6.
520. Marchetti, M., Tossani, N., Marchetti, S. and Bauce, G. (1999). Leaching of crystalline and coated vitamins in pelleted and extruded feeds. *Aquaculture*, 171(1–2):83–92. https://doi.org/10.1016/S0044–8486(98)00412–8.
521. Rumsey, G.L. (1991). Choline-betaine requirements of rainbow trout (*Oncorhynchus mykiss*). *Aquaculture*, 95(1–2):107–116. https://doi.org/10.1016/0044–8486(91)90077-K.
522. Nayak, S.K. (2010). Role of gastrointestinal microbiota in fish. *Aquaculture Research*, 41(11):1553–1573. https://doi.org/10.1111/j.1365–2109.2010.02546.x.
523. Hill, M.J. (1997). Intestinal flora and endogenous vitamin synthesis. *European Journal of Cancer Prevention: the Official Journal of the European Cancer Prevention Organisation (ECP)*, 6 (suppl. 1): S43–S45. https://doi.org/10.1097/00008469–199703001–00009.
524. Biesalski, H.K. (2016). Nutrition meets the microbiome: micronutrients and the microbiota. *Annals of the New York Academy of Sciences*, 1372, 1(1):53–64. https://doi.org/10.1111/nyas.13145.
525. Roem, A.J., Kohler, C.C. and Stickney, R.R. (1990). Inability to detect a choline requirement for the blue tilapia *Oreochromis aureus*. *Journal of the World Aquaculture Society*, 21(3):238–240. https://doi.org/10.1111/j.1749–7345.1990.tb01029.x.
526. Francis, G., Makkar, H.P. and Becker, K. (2001). Antinutritional factors present in plant-derived alternate fish feed ingredients and their effects in fish. *Aquaculture*, 199(3–4):197–227. https://doi.org/10.1016/S0044–8486(01)00526–9.
527. Prabu, E., Rajagopalsamy, C.B.T., Ahilan, B., Santhakumar, R., Jeevagan, I.J.M.A. and Renuhadevi, M. (2017). An overview of anti-nutritional factors in fish feed ingredients and their effects in fish. *Journal of Aquaculture in the Tropics*, 32(1/2):149–167.
528. Bora, P. (2014). Anti-nutritional factors in foods and their effects. *Journal of Academia and Industrial Research*, 3(6):285–290.

529. Krogdahl, Å., Penn, M., Thorsen, J., Refstie, S. and Bakke, A.M. (2010). Important antinutrients in plant feedstuffs for aquaculture: an update on recent findings regarding responses in salmonids. *Aquaculture Research*, 41(3):333–344. https://doi.org/10.1111/j.1365–2109.2009.02426.x.
530. Glencross, B.D., Baily, J., Berntssen, M.H., Hardy, R., MacKenzie, S. and Tocher, D.R. (2020). Risk assessment of the use of alternative animal and plant raw material resources in aquaculture feeds. *Reviews in Aquaculture*, 12(2):703–758. https://doi.org/10.1111/raq.12347.
531. Herbourne, J.B. (1989). Biosynthesis and functions of antinutritional factors in plants. *Aspects Appl. Biol.*, 19:21–28.
532. Tacon, A.G.J., Hasan, M.R. and Metian, M. (2011). Demand and supply of feed ingredients for farmed fish and crustaceans: trends and prospects. *FAO Fisheries and Aquaculture Technical Paper No. 564*. FAO, Rome.
533. Anderson, R.L. and Wolf, W.J. (1995). Compositional changes in trypsin inhibitors, phytic acid, saponins and isoflavones related to soybean processing. *Journal of Nutrition*, 125(3) (suppl. 3):581S–588S.
534. Van den Ingh, T.S.G.A.M., Krogdahl, Å., Olli, J.J., Hendriks, H.G.C.J.M. and Koninkx, J.G.J.F. (1991). Effects of soybean-containing diets on the proximal and distal intestine in Atlantic salmon (*Salmo salar*): a morphological study. *Aquaculture*, 94(4):297–305. https://doi.org/10.1016/0044–8486(91)90174–6.
535. Ayyagari, R., Rao, B.N. and Roy, D.N. (1989). Lectins, trypsin inhibitors, BOAA and tannins in legumes and cereals and the effects of processing. *Food Chemistry*, 34(3):229–238. https://doi.org/10.1016/0308–8146(89)90143-X.
536. Maenz, D.D., Irish, G.G. and Classen, H.L. (1999). Carbohydrate-binding and agglutinating lectins in raw and processed soybean meals. *Animal Feed Science and Technology*, 76(3–4):335–343. https://doi.org/10.1016/S0377–8401(98)00215–6.
537. Jenkins, K.J. and Atwal, A.S. (1994). Effects of dietary saponins on fecal bile acids and neutral sterols, and availability of vitamins A and E in the chick. *Journal of Nutritional Biochemistry*, 5(3):134–137. https://doi.org/10.1016/0955–2863(94)90084–1.
538. Knudsen, D., Røn, Ø., Baardsen, G., Smedsgaard, J., Koppe, W. and Frøkiær, H. (2006). Soyasaponins resist extrusion cooking and are not degraded during gut passage in Atlantic salmon (*Salmo salar* L.).

Journal of Agricultural and Food Chemistry, 54(17):6428–6435. https://doi.org/10.1021/jf0604992.

539. Iwashita, Y., Suzuki, N., Matsunari, H., Sugita, T. and Yamamoto, T. (2009). Influence of soya saponin, soya lectin, and cholyltaurine supplemented to a casein-based semipurified diet on intestinal morphology and biliary bile status in fingerling rainbow trout *Oncorhynchus mykiss*. *Fisheries Science*, 75(5):1307–1315. https://doi.org/10.1007/s12562-009-0158-1.
540. Couto, A., Kortner, T.M., Penn, M., Bakke, A.M., Krogdahl, Å. and Oliva-Teles, A. (2014). Effects of dietary phytosterols and soy saponins on growth, feed utilization efficiency and intestinal integrity of gilthead sea bream (*Sparus aurata*) juveniles. *Aquaculture*, 432: 295–303. https://doi.org/10.1016/j.aquaculture.2014.05.009.
541. Chen, W., Ai, Q., Mai, K., Xu, W., Liufu, Z., Zhang, W. and Cai, Y. (2011). Effects of dietary soybean saponins on feed intake, growth performance, digestibility and intestinal structure in juvenile Japanese flounder (*Paralichthys olivaceus*). *Aquaculture*, 318(1–2): 95–100. https://doi.org/10.1016/j.aquaculture.2011.04.050.
542. Cos, P., De Bruyne, T., Apers, S., Berghe, D.V., Pieters, L. and Vlietinck, A.J. (2003). Phytoestrogens: recent developments. *Planta Medica*, 69(07):589–599. https://doi.org/10.1055/s-2003-41122.
543. Cederroth, C.R. and Nef, S. (2009). Soy, phytoestrogens and metabolism: a review. *Molecular and Cellular Endocrinology*, 304(1–2): 30–42. https://doi.org/10.1016/j.mce.2009.02.027.
544. Pelissero, C., Bennetau, B., Babin, P., Le Menn, F. and Dunogues, J. (1991). The estrogenic activity of certain phytoestrogens in the Siberian sturgeon *Acipenser baeri*. *Journal of Steroid Biochemistry and Molecular Biology*, 38(3):293–299. https://doi.org/10.1016/0960-0760(91)90100-J.
545. El-Sayed, A.F.M., Abdel-Aziz, E.S.H. and Abdel-Ghani, H.M. (2012). Effects of phytoestrogens on sex reversal of Nile tilapia (*Oreochromis niloticus*) larvae fed diets treated with 17α-methyltestosterone. *Aquaculture*, 360:58–63. https://doi.org/10.1016/j.aquaculture.2012.07.010.
546. Gu, M., Gu, J.N., Penn, M., Bakke, A.M., Lein, I. and Krogdahl, Å. (2015). Effects of diet supplementation of soya-saponins, isoflavones and phytosterols on Atlantic salmon (*Salmo salar*, L) fry fed from start-feeding. *Aquaculture Nutrition*, 21(5):604–613. https://doi.org/10.1111/anu.12187.

547. Duncan, A.G. and Milne, J.A. (1989). Glucosinolates. *Aspects of Appl. Biol.*, 19:75–92.
548. Tripathi, M.K. and Mishra, A.S. (2007). Glucosinolates in animal nutrition. *Animal Feed Science and Technology*, 132(1–2):1–27. https://doi.org/10.1016/j.anifeedsci.2006.03.003.
549. Burel, C., Boujard, T., Escaffre, A.M., Kaushik, S.J., Boeuf, G., Mol, K.A., Van der Geyten, S. and Kuèhn, E.R. (2000). Dietary low-glucosinolate rapeseed meal affects thyroid status and nutrient utilization in rainbow trout (*Oncorhynchus mykiss*). *British Journal of Nutrition*, 83(6):653–664. https://doi.org/10.1017/S0007114500000830.
550. Hossain, M.A. and Jauncey, K. (1988). Toxic effects of glucosinolate (allyl isothiocyanate) (synthetic and from mustard oil cake) on growth and food utilization in common carp. *Indian Journal of Fisheries. Ernakulam*, 35(3):186–196.
551. Burel, C., Boujard, T., Kaushik, S.J., Boeuf, G., Van Der Geyten, S., Mol, K.A., Kühn, E.R., Quinsac, A., Krouti, M. and Ribaillier, D. (2000). Potential of plant-protein sources as fish meal substitutes in diets for turbot (*Psetta maxima*): growth, nutrient utilisation and thyroid status. *Aquaculture*, 188(3–4):363–382. https://doi.org/10.1016/S0044–8486(00)00342–2.
552. Davies, S.J., McConnell, S. and Bateson, R.I. (1990). Potential of rapeseed meal as an alternative protein source in complete diets for tilapia (*Oreochromis mossambicus* Peters). *Aquaculture*, 87(2):145–154. https://doi.org/10.1016/0044–8486(90)90271-N.
553. Enami, H.R. (2011). A review of using canola/rapeseed meal in aquaculture feeding. *Journal of Fisheries and Aquatic Science*, 6(1):22–36. https://doi.org/10.3923/jfas.2011.22.36.
554. Kumar, V., Sinha, A.K., Makkar, H.P., De Boeck, G. and Becker, K. (2012). Phytate and phytase in fish nutrition. *Journal of Animal Physiology and Animal Nutrition*, 96(3):335–364. https://doi.org/10.1111/j.1439–0396.2011.01169.x.
555. Lemos, D. and Tacon, A.G. (2017). Use of phytases in fish and shrimp feeds: a review. *Reviews in Aquaculture*, 9(3):266–282. https://doi.org/10.1111/raq.12138.
556. Dawood, M.A.O. and Koshio, S. (2020). Application of fermentation strategy in aquafeed for sustainable aquaculture. *Reviews in Aquaculture*, 12(2):987–1002. https://doi.org/10.1111/raq.12368.

557. Qiu, X. and Davis, D.A. (2017). Effects of dietary phytase supplementation on growth performance and apparent digestibility coefficients of Pacific white shrimp *Litopenaeus vannamei*. ***Aquaculture** Nutrition*, 23(5):942–951. https://doi.org/10.1111/anu.12462.
558. Hossain, M.A. and Jauncey, K. (1989). Nutritional evaluation of some Bangladeshi oilseed meals as partial substitutes for fish meal in the diet of common carp, *Cyprinus carpio* L. *Aquaculture Research*, 20(3): 255–268. https://doi.org/10.1111/j.1365–2109.1989.tb00351.x.
559. Ng, W.K. and Wee, K.L. (1989). The nutritive value of cassava leaf meal in pelleted feed for Nile tilapia. *Aquaculture*, 83(1–2):45–58. https://doi.org/10.1016/0044–8486(89)90059–8.
560. Lallès, J.P. and Peltre, G. (1996). Lead review article biochemical features of grain legume allergens in humans and animals. *Nutrition Reviews*, 54(4):101–107. https://doi.org/10.1111/j.1753–4887.1996.tb03883.x.
561. Kumar, S., Verma, A.K., Das, M. and Dwivedi, P.D. (2012). Allergenic diversity among plant and animal food proteins. *Food Reviews International*, 28(3):277–298. https://doi.org/10.1080/87559129.2011.635391.
562. Rumsey, G.L., Siwicki, A.K., Anderson, D.P. and Bowser, P.R. (1994). Effect of soybean protein on serological response, non-specific defense mechanisms, growth, and protein utilization in rainbow trout. *Veterinary Immunology and Immunopathology*, 41(3–4): 323–339. https://doi.org/10.1016/0165–2427(94)90105–8.
563. Kaushik, S.J., Cravedi, J.P., Lalles, J.P., Sumpter, J., Fauconneau, B. and Laroche, M. (1995). Partial or total replacement of fish meal by soybean protein on growth, protein utilization, potential estrogenic or antigenic effects, cholesterolemia and flesh quality in rainbow trout, *Oncorhynchus mykiss*. *Aquaculture*, 133(3–4):257–274. https://doi.org/10.1016/0044–8486(94)00403-B.
564. Seo, S.H. and Cho, S.J. (2016). Changes in allergenic and antinutritional protein profiles of soybean meal during solid-state fermentation with Bacillus subtilis. *LWT – Food Science and Technology*, 70:208–212. https://doi.org/10.1016/j.lwt.2016.02.035.
565. Refstie, S., Svihus, B., Shearer, K.D. and Storebakken, T. (1999). Nutrient digestibility in Atlantic salmon and broiler chickens related to viscosity and non-starch polysaccharide content in different soyabean products. *Animal Feed Science and Technology*, 79(4):331–345. https://doi.org/10.1016/S0377–8401(99)00026–7.

566. Aljuobori, A., Abdullah, N., Zulkifli, I., Soleimani, A.F., Liang, J.B. and Oskoueian, E. (2014). Lactobacillus salivarius fermentation reduced glucosinolate and fibre in canola meal. *Journal of Food Research*, 3(5):95–102. https://doi.org/10.5539/jfr.v3n5p95.
567. Castillo, S. and Gatlin 3rd, D.M. (2015). Dietary supplementation of exogenous carbohydrase enzymes in fish nutrition: a review. *Aquaculture*, 435:286–292. https://doi.org/10.1016/j.aquaculture.2014.10.011.
568. Li, M.H. and Robinson, E.H. (2006). Use of cottonseed meal in aquatic animal diets: a review. *North American Journal of Aquaculture*, 68(1):14–22. https://doi.org/10.1577/A05–028.1.
569. Hendricks, J.D. (2002). Adventitious toxins. In *Fish nutrition*, 3rd edition. J.E. Halver & R.W. Hardy (Eds.), pp. 601–649. Elsevier, Amsterdam, and Netherlands. https://doi.org/10.1016/B978–012319652–1/50011–2.
570. Bennett, J.W. and Klich, M. (2003). Mycotoxins. *Clinical Microbiology Reviews*, 16(3):497–516. https://doi.org/10.1128/CMR.16.3.497–516.2003.
571. Cinar, A. and Onbasi, E. (2019). Mycotoxins: the hidden danger in foods. *IntechOpen*. https://doi.org/10.5772/intechopen.89001.
572. Divakaran, S. and Tacon, A. (2000). Studies on the toxicity of shrimp (*Penaeus vannamei*) fed diets dosed with aflatoxin b1 to humans. *Journal of Aquatic Food Product Technology*, 9(3):115–120. https://doi.org/10.1300/J030v09n03_09.
573. NTP. (2013). June 2013. *ntp.niehs.nih.gov*. National Institute of Environmental Health Sciences, US Department of Health & Human Services, Mold.
574. Tola, M. and Kebede, B. (2016). Occurrence, importance and control of mycotoxins: a review. *Cogent Food and Agriculture*, 2(1). 1191103 https://doi.org/10.1080/23311932.2016.1191103.
575. Zain, M.E. (2011). Impact of mycotoxins on humans and animals. *Journal of the Saudi Chemical Society*, 15(2):129–144. https://doi.org/10.1016/j.jscs.2010.06.006.
576. Bryden, W.L. (2012). Mycotoxin contamination of the feed supply chain: implications for animal productivity and feed security. *Animal Feed Science and Technology*, 173(1–2):134–158. https://doi.org/10.1016/j.anifeedsci.2011.12.014.

577. Grenier, B. (2014). Effects of mycotoxins on the immune system. In *Sustain*:ability (E.M. Binder, Editor), World Nutrition Forum, Munich, October 15th–18th, 2014. Erber AG, pp. 327–336.
578. Liew, W.P.P. and Redzwan, M. (2018). Mycotoxins: its impact on gut health and microbiota. *Frontiers in Cellular and Infection Microbiology*, 8:60. Doi: 10.3389/fcimb.2018.00060 https://doi.org/10.3389/fcimb.2018.00060.
579. Pietsch, C. (2020). Risk assessment for mycotoxin contamination in fish feeds in Europe. *Mycotoxin Research*, 36(1):41–62. https://doi.org/10.1007/s12550–019–00368–6.
580. Eskola, M., Kos, G., Elliott, C.T., Hajšlová, J., Mayar, S. and Krska, R. (2019). Worldwide contamination of food-crops with mycotoxins: validity of the widely cited 'FAO estimate' of 25%. *Critical Reviews in Food Science and Nutrition*, DOI: 10.1080/10408398.2019.1658570 https://doi.org/10.1080/10408398.2019.1658570.
581. BIOMIN (2020). World mycotoxin survey. *Impact*, 2020. February 18th, 2020. https://www.biomin.net /science-hub/world-mycotoxin-survey-impact-2020/.
582. FAO. (2020). FishStatJ, a tool for fishery statistics analysis. *Release*. 4.00.10. Universal Software for Fishery Statistical Time Series. Global aquaculture production: quantity 1950–2018; Value 1950–2018; FAO, Rome.
583. Gonçalves, R.A., Schatzmayr, D., Hofstetter, U. and Stantos, G.A. (2017). Occurrence of mycotoxins in aquaculture: preliminary overview of Asian and European plant ingredients and finished feeds. *World Mycotoxin Journal*, 10(2):183–194.
584. Goncalves, R.A. and Muccio, M. (2019a). Fumonisins in aquaculture: occurrence, impact and counteraction. *Aqua Culture Asia Pacific Magazine*, 14(1):29–32.
585. Encarnacao, P., Santos, G. and Naehrer, K. (2014). Prevalence of mycotoxins in some major aquafeed ingredients a 2012 update. *Aqua Culture Asia Pacific Magazine*, 10(1):44–45.
586. Anater, A., Manyes, L., Meca, G., Ferrer, E., Luciano, F.B., Pimpao, C.T. and Font, G. (2016). Mycotoxins and their consequences in aquaculture: a review. *Aquaculture*, 451:1–10. https://doi.org/10.1016/j.aquaculture.2015.08.022.
587. Tacon, A.G.J. (2017). Biosecure shrimp feeds and feeding practices: guidelines for future development. *Journal of the World Aquaculture Society*, 48(3). https://doi.org/10.1111/jwas.12406.

588. Gonçalves, R.A., Naehrer, K. and Stantos, G.A. (2018a). Occurrence of mycotoxins in commercial aquafeeds in Asia and Europe: a real risk to aquaculture? *Reviews in Aquaculture*, 10(2):263–280. https://doi.org/10.1111/raq.12159.
589. Gonçalves, R.A., Tu, D.C., Nguyen, N.T., Santos, G.A., Encarnação, P. and Hung, L.T. (2018b). Aflatoxin b1 (AFB1) reduces growth performance, physiological response, and disease resistance in Tra catfish (*Pangasius hypophthalmus*). *Aquaculture International*, 26(3):921–936. https://doi.org/10.1007/s10499–018–0259-x.
590. Gonçalves, R.A., Schatzmayr, D., Albalat, A. and Mackenzie, S. (2020). Mycotoxins in aquaculture: feed and food. *Reviews in Aquaculture*, 12(1):145–175. https://doi.org/10.1111/raq.12310.
591. Nácher-Mestre, J., Beltrán, E., Strachan, F., Dick, J.R., Pérez-Sánchez, J., Berntssen, M.H.G. and Tocher, D.R. (2020). No transfer of the non-regulated mycotoxins, beauvericin and enniatins, from feeds to farmed fish reared on plant-based diets. *Food Chemistry*, 323: 1 September 2020, 126773 https://doi.org/10.1016/j.foodchem.2020.126773.
592. Jamshidizadeh, S., Narges, A.B., Yousefzadi, M. and Aramideh, A. (2019). Response of pacific white leg shrimp (*Litopenaeus vannamei*) on exposure to aflatoxin in feed. *Aquaculture Research*, 50(7): 1973–1984. https://doi.org/10.1111/are.14086.
593. Matejova, I., Svobodova, Z., Vakula, J., Mares, J. and Modra, H. (2017). Impact of mycotoxins on aquaculture fish species: a review. *Journal of the World Aquaculture Society*, 48(2):186–200. https://doi.org/10.1111/jwas.12371.
594. Ottinger, C.A. and Kaattari, S.L. (2000). Long-term immune dysfunction in rainbow trout (*Oncorhynchus mykiss*) exposed as embryos to aflatoxin b1. *Fish and Shellfish Immunology*, 10(1):101–106. https://doi.org/10.1006/fsim.1999.0227.
595. Rodriguez-Cervantes, C.H., Giron-Perez, M.I., Robledo-Marenco, M.L., Marin, S., Velazquez-Fernandez, J.B., Medina-Díaz, I., Rojas-García, A. and Ramos, A. (2010). Aflatoxin b1 and its toxic effects on immune response of teleost fishes: a review. *World Mycotoxin Journal*, 3(2):193–199. https://doi.org/10.3920/WMJ2009.1202.
596. Zeng, S.L., Long, W.Q., Tian, L.X., Xie, S.W., Chen, Y.J., Yang, H.J., Liang, G.Y. and Liu, Y.J. and Liu, Y.J. (2016). Effects of dietary aflatoxin b1 on growth performance, body composition,

haematological parameters and histopathology of juvenile pacific white shrimp (*Litopenaeus vannamei*). *Aquaculture Nutrition*, 22(5): 1152–1159. https://doi.org/10.1111/anu.12331.

597. Burgos-Hernandez, A., Farias, S.I., Torres-Arreola, W. and Ezquerra-Brauer, J.M. (2005). In vitro studies of the effects of aflatoxin b1 and fumonisin B1 on trypsin-like and collagenase-like activity from the hepatopancreas of white shrimp (*Litopenaeus vannamei*). *Aquaculture*, 250(1–2):399–410. https://doi.org/10.1016/j.aquaculture.2005.05.024.
598. García-Morales, M.-H., Pérez-Velázquez, M., González-Felix, M.L., Burgos-Hernández, A., Cortez-Rocha, M.-O., Bringas-Alvarado, L. and Ezquerra-Brauer, J. (2015). Effects of fumonisin-containing feed on the muscle proteins and ice-storage life of White shrimp (*Litopenaeus vannamei*). *Journal of Aquatic Food Product Technology*, 24(4):340. https://doi.org/10.1080/10498850.2013.775621.
599. Gbore, F.A., Adewole, A.M., Oginni, O., Oguntolu, M.F., Bada, A.M. and Akele, O. (2010). Growth performance, haematology and serum biochemistry of African catfish (*Clarias gariepinus*) fingerlings fed graded levels of dietary fumonisin B sub(1). *Mycotoxin Research*, 26(4):221–227. https://doi.org/10.1007/s12550–010–0059–2.
600. Tuan, N.A., Manning, B.B., Lovell, R.T. and Rottinghaus, G.E. (2003). Responses of Nile tilapia (*Oreochromis niloticus*) fed diets containing different concentrations of moniliformin or fumonisin B1. *Aquaculture*, 217(1–4):515–528. https://doi.org/10.1016/S0044–8486(02)00268–5.
601. Bernhoft, A., Høgåsen, H.R., Rosenlund, G., Moldal, T., Grove, S., Berntssen, M.H.G., Thoresen, S.I. and Alexander, J. (2018). Effects of dietary deoxynivalenol or ochratoxin A on performance and selected health indices in Atlantic salmon (*Salmo salar*). *Food and Chemical Toxicology*, 121:374–386. https://doi.org/10.1016/j.fct.2018.08.079.
602. Diab, A.M.R.M., Salem, E.-K., Ali, M.S., Gehan, I.E. and El-Habashi, N. (2019). Experimental ochratoxicosis A in Nile tilapia and its amelioration by some feed additives. *International Journal of Veterinary Sciences*, 6(2):149–158. https://doi.org/10.1016/j.ijvsm.2018.09.004.
603. Manning, B.B., Ulloa, R.M., Li, M.H., Robinson, E.H. and Rottinghaus, G.E. (2003a). Ochratoxin A fed to channel catfish (*Ictalurus punctatus*) causes reduced growth and lesions of

hepatopancreatic tissue. *Aquaculture*, 219(1–4):739–750. https://doi.org/10.1016/S0044–8486(03)00033–4.

604. Manning, B.B., Terhune, J.S., Li, M.H., Robinson, E.H., Wise, D.J. and Rottinghaus, G.E. (2005). Exposure to feedborne mycotoxins T-2 toxin or ochratoxin A causes increased mortality of Channel catfish challenged with *Edwardsiella ictaluri*. *Journal of Aquatic Animal Health*, 17(2):147–152. https://doi.org/10.1577/H03–063.1.
605. Ghareeb, K., Awad, W.A., Bohm, J. and Zebeli, Q. (2015). Impacts of the feed contaminant deoxynivalenol on the intestine of monogastric animals: poultry and swine. *Journal of Applied Toxicology*, 35(4):327–337. https://doi.org/10.1002/jat.3083.
606. Gonçalves, R.A., Menanteau-Ledouble, S., Schöller, M., Eder, A., Schmidt-Posthaus, H., Mackenzie, S. and El-Matbouli, M. (2019c). Effects of deoxynivalenol exposure time and contamination levels on rainbow trout. *Journal of the World Aquaculture Society*, 50(1):137–154. https://doi.org/10.1111/jwas.12542.
607. Hooft, J.M., Elmor, A.-H., Encarnacao, P. and Bureau, D.P. (2011). Rainbow trout (*Oncorhynchus mykiss*) is extremely sensitive to the feed-borne Fusarium mycotoxin deoxynivalenol (DON). *Aquaculture*, 311(1–4):224–232. https://doi.org/10.1016/j.aquaculture.2010.11.049.
608. Manning, B.B., Abbas, H.K., Wise, D.J. and Greenway, T. (2014). The effect of feeding diets containing deoxynivalenol contaminated corn on channel catfish (*Ictalurus punctatus*) challenged with *Edwardsiella ictalurid*. *Aquaculture Research*, 45(11):1782–1786. https://doi.org/10.1111/are.12123.
609. Pietsch, C., Michel, C., Kersten, S., Valenta, H., Daenicke, S., Schulz, C., Kloas, W. and Burkhardt-Holm, P. (2014). In vivo effects of deoxynivalenol (DON) on innate immune responses of carp (*Cyprinus carpio* L.). *Food and Chemical Toxicology*, 68:44–52. https://doi.org/10.1016/j.fct.2014.03.012.
610. Ryerse, I.A., Hooft, J.M., Bureau, D.P., Hayes, M.A. and Lumsden, J.S. (2015). Purified deoxynivalenol or feed restriction reduces mortality in rainbow trout, *Oncorhynchus mykiss* (Walbaum), with experimental bacterial coldwater disease but biologically relevant concentrations of deoxynivalenol do not impair the growth of Flavobacterium psychrophilum. *Journal of Fish Diseases*, 38(9):809–819. https://doi.org/10.1111/jfd.12295.

611. Ryerse, I.A., Hooft, J.M., Bureau, D.P., Hayes, A., M. and Lumsden, J.S. (2016). Diets containing corn naturally contaminated with deoxynivalenol reduces the susceptibility of rainbow trout (*Oncorhynchus mykiss*) to experimental Flavobacterium psychrophilum infection. *Aquaculture Research*, 47(3):787–796. https://doi.org/10.1111/are.12537.
612. Kravchenko, L.V., Galash, V.T., Avren'eva, L.T. and Kranauskas, A.E. (1989). On the sensitivity of carp, *Cyprinus carpio*, to mycotoxin T-2. *Journal of Ichthyology*, 29(7):156–160.
613. Modra, H., Sisperova, E., Blahova, J., Enevova, V., Fictum, P., Franc, A., Mares, J. and Svobodova, Z. (2018). Elevated concentrations of T-2 toxin cause oxidative stress in the rainbow trout (*Oncorhynchus mykiss*). *Aquaculture Nutrition*, 24(2):842–849. https://doi.org/10.1111/anu.12613.
614. Manning, B.B., Li, M.H., Robinson, E.H., Gaunt, P.S., Camus, A.C. and Rottinghaus, G.E. (2003b). Response of channel catfish to diets containing T-2 toxin. *Journal of Aquatic Animal Health*, 15(3):229–238. https://doi.org/10.1577/H03–019.
615. Abdel-Tawwab, M., Khalifa, E., Diab, A.M., Khallaf, M.,A. and Abdel-Razek, N. (2020). Dietary garlic and chitosan alleviated zearalenone toxic effects on performance, immunity, and challenge of European sea bass, *Dicentrarchus labrax*, to *Vibrio alginolyticus* infection. *Aquaculture International*, 28(2):493–510. https://doi.org/10.1007/s10499–019–00477–0.
616. Prearo, M., Minervini, F., Pavoletti, E., Pascale, M., Visconti, A.; et al. (2000). Symtomatological, anatomopathological and histological features in rainbow trout fed with a deoxynivalenol (vomitoxin) contaminated diet. *Bollettino Societa Italiana di Patologia Ittica*, 12(29):33–43.
617. Wony, M., Dobosz, S., Obremski, K., Hliwa, P., Gomuka, P., Łakomiak, A., Różyński, R., Zalewski, T. and Brzuzan, P. (2015). Feed-borne exposure to zearalenone leads to advanced ovarian development and limited histopathological changes in the liver of premarket size rainbow trout. *Aquaculture*, 448:71–81. https://doi.org/10.1016/j.aquaculture.2015.05.032.
618. Wozny, M., Obremski, K., Jakimiuk, E., Gusiatin, M. and Brzuzan, P. (2013). Zearalenone contamination in rainbow trout farms in North-Eastern Poland. *Aquaculture*, 416–417:209. https://doi.org/10.1016/j.aquaculture.2013.09.030.

619. Goncalves, R.A. and Muccio, M. (2019b). Current knowledge about fumonisins in aquaculture, 4 March 2019. Global Aquaculture Advocate. https://www.aquaculturealliance.org/current-knowledge-about-fumonisins-aquaculture.
620. Boonyaratpalin, M., Supamattaya, K., Verakunpiriya, V. and Suprasert, D. (2001). Effects of aflatoxin b1 on growth performance, blood components, immune function and histopathological changes in black tiger shrimp (*Penaeus monodon fabricius*). *Aquaculture Research*, 32:388–398. https://doi.org/10.1046/j.1355–557x.2001.00046.x.
621. Huang, Q., Tan, B., Dong, X., Zhang, S., Yang, Q. and Chi, S. (2018). Effect of chitosan oligosaccharide and mycotoxin adsorbents on intestinal mucosa structure and gut flora of *Litopenaeus vannamei.* Zhongguo Shui Chan Ke Xue, Beijing, 2(2):373–383. https://doi.org/10.3724/SP.J.1118.2018.17236.
622. Hasan, A. and Goncalves, R.A. (2017). Aflatoxin: A serious immunosuppressant in aquaculture. *Aqua Culture Asia Pacific Magazine*, July/August:49–51.
623. Romer Labs. (2016). Worldwide mycotoxin regulations. https://www.romerlabs.com/en/knowledge-center/knowledge-library/articles/news/worldwide-mycotoxin-regulations/.
624. Pierron, A., Alassane-Kpemb, I. and Oswalda, I.P. (2016). Impact of mycotoxin on immune response and consequences for pig health. *Animal Nutrition*, 2(2):63–68. https://doi.org/10.1016/j.aninu.2016.03.001.
625. Turner, N.W., Subrahmanyam, S. and Piletsky, S.A. (2009). Analytical methods for determination of mycotoxins: a review. *Analytica Chimica Acta*, 632(2):168–180. https://doi.org/10.1016/j.aca.2008.11.010.
626. Zheng, M.Z., Richard, J.L. and Binder, J. (2006). A review of rapid methods for the analysis of mycotoxins. *Mycopathologia*, 161(5):261–273. https://doi.org/10.1007/s11046–006–0215–6.
627. Dwivedy, A.K., Kumar, M., Upadhyay, N., Prakash, B. and Dubey, N.K. (2016). Plant essential oils against food borne fungi and mycotoxins. *Current Opinion in Food Science*, 11:16–21. https://doi.org/10.1016/j.cofs.2016.08.010.
628. Hassan, R., El-Kadi, S. and Mostafa, S. (2015). Effect of some organic acids on some fungal growth and their toxins. *International Journal of Advances in Biology*, 2(1):1–11.

629. Higgins, C. and Brinkhaus, F. (1999). Efficacy of several organic acids against molds. *Journal of Applied Poultry Research*, 8(4): 480–487. https://doi.org/10.1093/japr/8.4.480.
630. Ray, L.L. and Bullerman, L.B. (1982). Preventing growth of potentially toxic molds using antifungal agents. *Journal of Food Protection*, 45(10):953–963. https://doi.org/10.4315/0362–028X-45.10.953.
631. Ricke, S.C. (2003). Perspectives on the use of organic acids and short chain fatty acids as antimicrobials. *Poultry Science*, 82(4):632–639. https://doi.org/10.1093/ps/82.4.632.
632. Theron, M.M. and Rykers Lues, J.F. (2010). *Organic acids and food preservation*. CRC Press, Boca Raton, FL. https://doi.org/10.1201/9781420078435.
633. Avantaggiato, G., Solfrizzo, M. and Visconti, A. (2005). Recent advances on the use of adsorbent materials for detoxification of Fusarium mycotoxins. *Food Additives and Contaminants*, 22(4): 379–388. https://doi.org/10.1080/02652030500058312.
634. Garcia-Perez, O., Tapia-Salazar, M., Nieto-Lopez, M., Villarreal-Cavazos, D., Cruz-Suarez, L. and Ricque-Marie, D. (2013). Effectiveness of aluminosilicate-based products for detoxification of aflatoxin-contaminated diets for juvenile pacific white shrimp, *Litopenaeus vannamei*. *Ciencias Marinas*, 39(1):1–13. https://doi.org/10.7773/cm.v39i1.2165.
635. Hassan, M.S., Nasar, K.M., Mohammady, E.Y., Amin, A., Tayel, S.I. and El-Haroun, E.R. (2020). Nano-zeolite efficiency to mitigate the aflatoxin b1 (AFB1) toxicity: effects on growth, digestive enzymes, antioxidant, DNA damage and bioaccumulation of AFB1 residues in Nile tilapia (*Oreochromis niloticus*). *Aquaculture*, 523(June):article 735123. https://doi.org/10.1016/j.aquaculture.2020.735123.
636. Kumar, V., Roy, S., Barman, D., Kumar, A., Powl, L. and W.A. Meeti. (2013). Importance of mycotoxins in aquaculture feeds. *Aquaculture Asia Pacific Magazine*, Jan/Mar:25–29.
637. Leeson, S., Diaz, G.J. and Summers, J.D. (1995). *Poultry metabolic disorders and mycotoxins*. University Books, Guelph, Ontario.
638. Louyer, A. and Gallissot, M. (2013). Effects of mycotoxin binder on male red tilapia performance. *Aquaculture Asia Pacific Magazine*, May/June:20–21.
639. Mitchell, C. and Sinnott, R. (2014). The unseen enemy. *International Aquafeed*, Sept/Oct:12.

640. Pimpukdee, K. (2004). Adsorption of aflatoxin b1 by a variety of adsorbents. *International Aquafeed*, Mar:30–31.
641. Ratcliff, J. (2005). The role of mycotoxins in food and feed safety. *International Aquafeed*, July/Aug 2005:14–18.
642. Sealey, W.M., Hooley, C.G., Rosentrater, K.A., Gaylord, T.G. and Barrows, F.T. (2015). The effect of a mycotoxin deactivation product on growth of juvenile rainbow trout fed distillers dried grains. *North American Journal of Aquaculture*, 77(4):429–436. https://doi.org/10.1080/15222055.2015.1029175.
643. Tapia-Salazar, M., García-Pérez, O.,D., Nieto-López, M.,G., Villarreal-Cavazos, D., Gamboa-Delgado, J., Cruz-Suárez, L.E. and Ricque-Marie, D. (2017). Evaluating the efficacy of commercially available aflatoxin binders for decreasing the effects of aflatoxicosis on pacific white shrimp *litopenaeus vannamei*. *Hidrobiologica*, 27(3): 411–418. https://doi.org/10.24275/uam/izt/dcbs/hidro/2017v27n3/Tapia.
644. Van Schoonhoven, M.J. and Gallissot, M. (2017). Mycotoxin detoxification: science vs marketing. *Aquaculture Asia Pacific Magazine*, September/October, 17:41–43.
645. Guan, S., He, J., Young, J.C., Zhu, H., Li, X.Z., Ji, C. and Zhou, T. (2009). Transformation of trichothecene mycotoxins by microorganisms from fish digesta. *Aquaculture*, 290(3–4):290–295. https://doi.org/10.1016/j.aquaculture.2009.02.037.
646. Jard, G., Liboz, T., Mathieu, F., Guyonvarc'h, A. and Lebrihi, A. (2011). Review of mycotoxin reduction in food and feed: from prevention in the field to detoxification by adsorption or transformation. *Food Additives and Contaminants, Part A*, 28(11):1590–1609. https://doi.org/10.1080/19440049.2011.595377.
647. Loi, M., Fanelli, F., Liuzzi, V.C., Logrieco, A.F. and Mulè, G. (2017). Mycotoxin biotransformation by native and commercial enzymes: present and future perspectives. *Toxins*, 9(4):111. doi:10.3390/toxins9040111 https://doi.org/10.3390/toxins9040111.
648. McCormick, S.P. (2013). Microbial detoxification of mycotoxins. *Journal of Chemical Ecology*, 39(7):907–918. https://doi.org/10.1007/s10886–013–0321–0.
649. Adebiyi, J.A., Kayitesi, E., Adebo, O.A., Changwa, R. and Njobeh, P.B. (2019). Food fermentation and mycotoxin detoxification: an African perspective. *Food Control*, 106. 106731731 https://doi.org/10.1016/j.foodcont.2019.106731.

650. Bata, Á. and Lásztity, R. (1999). Detoxification of mycotoxin-contaminated food and feed by microorganisms. *Trends in Food Science and Technology*, 10(6–7):223–228. https://doi.org/10.1016/S0924–2244(99)00050–3.
651. Hathout, A.S. and Aly, S.E. (2014). Biological detoxification of mycotoxins: a review. *Annals of Microbiology*, 64(3):905–919. https://doi.org/10.1007/s13213–014–0899–7.
652. Juodeikiene, G., Basinskiene, L., Bartkiene, E. and Matusevicius, P. (2012). Mycotoxin decontamination aspects in food, feed and renewables using fermentation processes. *Structure and Function of Food Engineering*, 2012(August 22nd):171–204. https://doi.org/10.5772/46184.
653. GAA. (2016). Best aquaculture practices – fact sheet. https://www.aquaculture alliance.org/wp-content/uploads/2016/03/ BAP-Fact-Sheet-LT.pdf.
654. FAO. (2001). Good aquaculture feed manufacturing practice. *FAO Technical Guidelines for Responsible Fisheries*. 5 (suppl. 1). FAO, Rome.
655. Thompson, L.A. and Darwish, W.S. (2019). Environmental chemical contaminants in food: review of a global problem. *Journal of Toxicology*. https://doi.org/10.1155/2019/2345283.
656. Berntssen, M.H.G., Maage, A., Julshamn, K., Oeye, B.E. and Lundebye, A.-K. (2011). Carry-over of dietary organochlorine pesticides, PCDD/Fs, PCbs, and brominated flame retardants to Atlantic salmon (*Salmo salar* L.) fillets. *Chemosphere*, 83(2):95–103.
657. Hellberg, R.S., De Witt, C.A.M. and Morrissey, M.T. (2012). Risk-benefit analysis of seafood consumption: a review. *Compr. Rev. Food Sci.* F, 11(5):490–517.
658. Vitenskapskomiteen, V.K.M. (2014). *Benefit-risk assessment of fish and fish products in the Norwegian diet – an update. Scientific Opinion of the scientific Steering Committee.* VKM Report 15. Oslo, Norway.
659. Xanthos, D. and Walker, T.R. (2017). International policies to reduce plastic marine pollution from single-use plastics (plastic bags and microbeads): a review. *Marine Pollution Bulletin*, 118(1–2): 17–26. https://doi.org/10.1016/j.marpolbul.2017.02.048.
660. Bøhn, T., Cuhra, M., Traavik, T., Sanden, M., Fagan, J. and Primicerio, R. (2014). Compositional differences in soybeans on the market: glyphosate accumulates in Roundup Ready GM soybeans. *Food Chemistry*, 153:207–215. https://doi.org/10.1016/j.foodchem.2013.12.054.

661. Krüger, M., Schledorn, P., Schrödl, W., Hoppe, H.W., Lutz, W. and Shehata, A.A. (2014). Detection of glyphosate residues in animals and humans. *Journal of Environmental and Analytical Toxicology*, 4(2):1–5.
662. Myers, J.P., Antoniou, M.N., Blumberg, B., Carroll, L., Colborn, T. et al. (2016). Concerns over use of glyphosate-based herbicides and risks associated with exposures: a consensus statement. *Environmental Health*, 15(1):1–13. https://doi.org/10.1186/s12940–016–0117–0.
663. Cuhra, M., Bøhn, T. and Cuhra, P. (2016). Glyphosate: too much of a good thing? *Frontiers in Environmental Science*, 4:28. https://doi.org/10.3389/fenvs.2016.00028.
664. Kang, H.S., Lee, S.B., Shin, D., Jeong, J., Hong, J.H. and Rhee, G.S. (2018). Occurrence of veterinary drug residues in farmed fishery products in South Korea. *Food Control*, 85:57–65. https://doi.org/10.1016/j.foodcont.2017.09.019.
665. Kim, J., Suh, J.H., Cho, H.D., Kang, W., Choi, Y.S. and Han, S.B. (2016). Analytical method for fast screening and confirmation of multi-class veterinary drug residues in fish and shrimp by LC-MS/MS. *Food Additives and Contaminants: Part A*, 33(3):420–432. https://doi.org/10.1080/19440049.2016.1139752.
666. Lunestad, B.T. and Samuelsen, O. (2008). Veterinary drug use in aquaculture. In *Improving farmed fish quality and safety*, pp. 97–127. Woodhead Publishing. https://doi.org/10.1533/9781845694920.1.97.
667. Santos, L. and Ramos, F. (2016). Analytical strategies for the detection and quantification of antibiotic residues in aquaculture fishes: a review. *Trends in Food Science and Technology*, 52:16–30. https://doi.org/10.1016/j.tifs.2016.03.015.
668. Stolker, A.A.M., Zuidema, T., Nielen, M.W.F. and Nielen, M.W.F. (2007). Residue analysis of veterinary drugs and growth-promoting agents. *TrAC Trends in Analytical Chemistry*, 26(10):967–979. https://doi.org/10.1016/j.trac.2007.09.008.
669. Lightner, D.V., Pantoja, C.R., Redman, R.M., Hasson, K.W. and Menon, J.P. (2009). Case reports of melamine-induced pathology in penaeid shrimp fed adulterated feeds. *Diseases of Aquatic Organisms*, 86(2):107–112. https://doi.org/10.3354/dao02111.
670. Karunasagar, I. (2009). Melamine in fish feed and implications for safety of aquaculture products. *FAO Aquaculture Newsletter*, 42:29–31.

671. Phromkunthong, W., Choochuay, P., Kiron, V., Nuntapong, N. and Boonyaratpalin, M. (2015). Pathophysiological changes associated with dietary melamine and cyanuric acid toxicity in red tilapia. *Journal of Fish Diseases*, 38(2):161–173. https://doi.org/10.1111/jfd.12219.
672. Bolden, A.L., Rochester, J.R. and Kwiatkowski, C.F. (2017). Melamine, beyond the kidney: A ubiquitous endocrine disruptor and neurotoxicant? *Toxicology Letters*, 280:181–189. https://doi.org/10.1016/j.toxlet.2017.07.893.
673. Matthews, D. and Cooke, B.C. (2003). The potential for transmissible spongiform encephalopathies in non-ruminant livestock and fish. *Revue Scientifique et Technique-de l'Office International des Epizooties*, 22(1):283–296. https://doi.org/10.20506/rst.22.1.1393.
674. Dalla Valle, A.Z., Iriti, M., Faoro, F., Berti, C. and Ciappellano, S. (2008). In vivo prion protein intestinal uptake in fish. *APMIS*, 116(3):173–180. https://doi.org/10.1111/j.1600–0463.2008.00863.x.
675. Lunestad, B.T. and Rosnes, J.T. (2008). Microbiological quality and safety of farmed fish. In *Improving farmed fish quality and safety*, pp. 399–427. Woodhead Publishing. https://doi.org/10.1533/9781845694920.2.399.
676. Salta, E., Panagiotidis, C., Teliousis, K., Petrakis, S., Eleftheriadis, E., Arapoglou, F., Grigoriadis, N., Nicolaou, A., Kaldrymidou, E., Krey, G. and Sklaviadis, T. (2009). Evaluation of the possible transmission of BSE and scrapie to gilthead sea bream (*Sparus aurata*). *PLOS ONE*, 4(7). https://doi.org/10.1371/journal.pone.0006175.
677. Bakke-McKellep, A.M., Koppang, E.O., Gunnes, G., Sanden, M., Hemre, G.I., Landsverk, T. and Krogdahl, Å. (2007). Histological, digestive, metabolic, hormonal and some immune factor responses in Atlantic salmon, *Salmo salar* L., fed genetically modified soybeans. *Journal of Fish Diseases*, 30(2):65–79. https://doi.org/10.1111/j.1365–2761.2007.00782.x.
678. Hemre, G.I., Sanden, M., Bakke-Mckellep, A.M., Sagstad, A. and Krogdahl, Å. (2005). Growth, feed utilization and health of Atlantic salmon *Salmo salar* L. fed genetically modified compared to non-modified commercial hybrid soybeans. *Aquaculture Nutrition*, 11(3):157–167. https://doi.org/10.1111/j.1365–2095.2005.00328.x.
679. Wiik-Nielsen, C.R., Holst-Jensen, A., Bøydler, C. and Berdal, K.G. (2011). Quantification of dietary DNA in tissues of Atlantic

salmon (*Salmo salar* L.) fed genetically modified feed ingredients. *Aquaculture Nutrition*, 17(2):e668-e674. https://doi.org/10.1111/j.1365–2095.2010.00817.x.

680. Dona, A. and Arvanitoyannis, I.S. (2009). Health risks of genetically modified foods. *Critical Reviews in Food Science and Nutrition*, 49(2):164–175. https://doi.org/10.1080/10408390701855993.
681. Sørensen, M., Berge, G.M., Thomassen, M.S., Ruyter, B., Hatlen, B., Ytrestøyl, T., Aas, T.S. and Åsgård, T.E. (2011). Today's and tomorrow's feed ingredients in Norwegian aquaculture. *Nofima Report Series, Report 52/2011.*
682. Love, D.C., Halden, R.U., Davis, M.F. and Nachman, K.E. (2012). Feather meal: a previously unrecognized route for reentry into the food supply of multiple pharmaceuticals and personal care products (PPCPs). *Environmental Science and Technology*, 46(7):3795–3802. https://doi.org/10.1021/es203970e.
683. Krogdahl, A., Kortner, T.M. and Hardy, R.W. (2021). Antinutrients and adventitious toxins. In *Fish nutrition*, 4th edition. R.W. Hardy & S.J. Kaushik (Eds.), pp. 775–821, 922p. Academic Press, Elsevier Inc., London ISBN: 978–0-12–819587–1. https://doi.org/10.1016/B978–0-12–819587–1.00001-X.

Index

Explanatory note: The letter t following a page number denotes a table; the letter f following a page number denotes a figure.